D0618090

SUBMARINE

ACTION

SUBMARINE
ACTION

PAUL KEMP

CHANCELLOR PRESS

First published in 1999 by
Sutton Publishing Limited · Phoenix Mill
Thrupp · Stroud · Gloucestershire · GL5 2BU

This edition published in 2000 by Chancellor Press,
an imprint of Bounty Books,
a division of Octopus Publishing Group Limited,
2–4 Heron Quays, London, E14 4JP.

British Library Cataloguing in Publication Data
A catalogue record for this book is available from the British Library

ISBN 0-75370-334-3

Typeset in 10/13pt Sabon.
Typesetting and origination by
Sutton Publishing Limited.
Printed in Great Britain by
J.H. Haynes & Co., Sparkford.

CONTENTS

ACKNOWLEDGEMENTS

I am extremely grateful to the following for their help, in so many different ways, in the preparation of this book: the Director and Staff of the Black Sea Fleet Museum in Sevastopol and their Bulgarian counterparts at Varna; Richard J. Boyle; Horst Bredow of the *U-boot Archiv*; the late and great Gus Britton of the Royal Navy Submarine Museum; Cdr John Bull RN; Ian Buxton; J.K. Chapman; the late Capt John Coote RN; Debbie Corner of the Royal Navy Submarine Museum; George Cuddon; Charles Deleay; Lt Cdr R.B. Foster RN; Brian Head; David Hill; *Capitaine de Vauisseau* Claude Huan; R.G. Jones; the Historical Service of the Royal Netherlands Navy; Dr Peter Jung of the Kriegsarchiv in Vienna; the staff of the superb Kent County Council Reference Library at Springfield in Maidstone, especially Lynne Burroughs who made countless ascents of the 'tower' in search of material; the late Vice Adm Sir Hugh Mackenzie; the staff of the National Maritime Museum, Greenwich; the late Cdr Arthur Pitt RN; Judith Prendergast; the staff of the Public Record Office in London; Dott. Achille Rastelli; Lt Cdr R.G. Raikes; Vice Adm Sir John Roxburgh; the Royal Swedish Navy; the *Service Historique de la Marine* in Paris; United States National Archives; United States Naval Institute and Wright & Logan. The authorities in the Democratic Peoples' Republic of North Korea and the Islamic Republic of Iran are still unable to assist me with enquiries relating to submarines employed by those nations. Lastly, but no means least, I must thank Jonathan Falconer at Sutton Publishing for taking the idea onboard and Alison Flowers for editing and making sense of the manuscript. Finally I must acknowledge the unseen role played in all this by the officers and men of HMS *Finwhale* who in 1974 took a fifteen-year-old schoolboy to sea and, unbeknown to them, started a love affair with the submarine which endures to this day.

GLOSSARY

AA	Anti-Aircraft
ABM	Anti-Ballistic Missile
AEW	Airborne Early Warning
AIP	Air Independent Propulsion
AIV	Automatic Inboard Venting. A system by which the air used to fire a torpedo is drawn back into the submarine rather than released from the torpedo tube and thus giving the submarine's position away.
ASV	Air to Surface Vessel. Designation for airborne radar during the Second World War.
ASW	Anti-Submarine Warfare
BdU	*Befehlshaber der U-boote* (Commander-in-Chief U-boats). This was the appointment held by Karl Dönitz from October 1939 until January 1943.
Bold	German Second World War designation for accoustic decoy released from a submarine to produce a false echo on an attacking ship's sonar.
CAT	Canadian-designed anti-acoustic torpedo apparatus
CC	*Capitano di Corvetta* (Italian Navy: Lieutenant-Commander)
C-in-C	Commander-in-Chief
CO	Commanding Officer
COPP	Combined Operations Pilotage Parties: British wartime organisation responsible for covert reconnaissance of likely invasion beaches
CSLBM	Conventionally Armed Submarine Launched Ballistic Missile
CVE	Escort or auxiliary aircraft carrier
D/F	Direction Finding
DNI	Director of Naval Intelligence (British)
Dräger	German submarine escape apparatus very similar to British DSEA
DSEA	Davis Submarine Escape Apparatus
DSMAC	Digital Scene Matching Area Correlator
DSRV	Deep Submergence Rescue Vehicle
ECE	External Combustion Engine
ECM	Electronic Counter-Measures
EG	Escort Group
ELF	Extremely Low Frequency
End Around	High-speed surface manoeuvre designed to place the

	submarine ahead of an advancing convoy. The submarine then dives and prepares to carry out a dived attack. It was much practised by US submarines in the Pacific during the Second World War – their high surface speed allowed them to engage in such manoeuvres.
ESM	Electronic Support Measures
FAC	Fast Attack Craft
FdU	*Führer der U-boote* (Senior Officer U-boats). This was the appointment held by Karl Dönitz before his promotion to flag rank in 1939. It could also be used in a local context; for example, *FdU Mittelmeer*, Senior Officer U-boats, Mediterranean.
Fido	US-designed acoustic torpedo dropped by aircraft for use against submarines. For security reasons the weapon was known as the Mk.XXIV mine.
FOSM	Flag Officer Submarines
Foxer	British and American anti-acoustic torpedo apparatus
Gash	British term for rubbish
GIUK	Greenland–Iceland–UK line. Postwar NATO acronym for the sea passages between these islands through which Soviet submarines would have to pass on their way into the Atlantic.
GNAT	German Naval Acoustic Torpedo. British term for the German T5 *Zaunkonig* acoustic homing torpedo.
HE	Hydrophone Effect
Heads	British and American term for toilets
HF	High Frequency
HF/DF	High Frequency Direction Finding, sometimes referred to as 'Huff Duff'
HP	High Pressure
ICE	Internal Combustion Engine
JATO	Jet Assisted Take-Off
KL	*Kapitänleutnant* (German Navy: Lieutenant Commander)
LF	Low Frequency
LI	*Leitender Ingenieur* (Engineer Officer in a U-boat)
MAC	Merchant Aircraft Carrier. These were bulk grain carriers or oil tankers with the superstructure removed and replaced by a flight-deck; some had limited hangar space. They successfully combined the functions of merchant ship and aircraft carrier without detriment to either.
MAD	Magnetic Anomaly Detection
MBT	Main Ballast Tanks
MIRV	Multiple Independently Targeted Re-Entry Vehicle. The warhead on a missile fired by an SSBN which contains a number of nuclear weapons each assigned to its own target. These warheads are designed to confuse ABM defences.
MPA	Maritime Patrol Aircraft

MTM	Main Tactical Mast
NDB	Nuclear Depth Bomb
OKW	*Oberkommando des Wehrmacht*. The German High Command during the Second World War.
OMC	One-Man Control
OMEC	One-Man Escape Chamber (British)
OOW	Officer of the Watch
OSS	Office of Strategic Services
PDM	Passive Delay Mechanism. Device fitted to ground mines which allows the mine to remain inert for a pre-set time or until a number of ships have passed over the mine.
PLARB	Soviet designation for an SSBN
PLE	Prudent Limit of Endurance. The time on patrol when an anti-submarine aircraft has sufficient fuel remaining, and no more, for the return flight back to base.
PWR	Pressurised Water Reactor
RA(S)	See VA(S)
ROE	Rules of Engagement
Run Hot	A situation where the engine of a torpedo starts to run while the weapon is still sealed in the torpedo tube or stowed in the fore ends. The engine produces dangerous amounts of gas which can cause an explosion.
SAM	Surface to Air Missile
SBT	Submarine Bubble Target
SEALs	Postwar US Special Forces
SEIS	Submarine Escape Immersion Suit (British)
SETT	Submarine Escape Training Tank (British)
SDV	Swimmer Delivery Vehicle
SKL	*Seekriegsleitung* (German Naval High Command)
SLAM	Submarine Launched Airflight Missile
SLCM	Submarine Launched Cruise Missile
SOE	Senior Officer of the Escort or Special Operations Executive
SSBN	Ballistic missile submarine
SSGN	Nuclear-powered fleet submarine armed with SLCMs
SSK	Diesel-electric-powered submarine
SSN	Nuclear-powered fleet submarine
SSR	Submarine Radar Picket, USN classification
TAINS	TERCOM-Aided Inertial Navigation System
TERCOM	Terrain Contour Matching
TGM	Torpedo Gunner's Mate
TMD	Tube Mounted Dispenser
TV	*Tenente de Vascello* (Italian Navy: Lieutenant)
'Uckers'	Game similar to ludo played with great passion by British submariners
UDT	Underwater Demolition Teams: US Navy wartime equivalent of COPP but also specialising in demolition of underwater obstacles.
UQS	Ultra Quiet State

USN	United States Navy
USNR	United States Navy Reserve
VA(S)	Vice-Admiral (Submarines) the head of the Royal Navy's Submarine Service. The title could also be RA(S) if the officer concerned was a Rear Admiral. In 1942 the title changed to Flag Officer Submarines, FOSM, and has remained thus to this day.
VLF	Very Low Frequency
WRNS	Women's Royal Naval Service (British)
XO	Executive Officer (US Navy)
XBT	Expendable Bathythermograph
Zampolit	Political officer in the former Soviet Navy

WAR BENEATH THE SEA

What will become of maritime wars, and where will sailors be found to man ships of war, when it is a physical certainty that they may every moment be blown into the air by means of a diving boat, against which no human foresight can guard them?

Robert Fulton

This book is about change. It is also about what has remained the same. It is about the history of the submarine in twentieth-century naval operations, and about the design, construction and propulsion of submarines. It is about the men who live and work in the vessels and about the particular varieties of undersea operations. The twentieth century has seen the rise and fall of many types of warship: the battleship, the battle-cruiser and the aircraft carrier to name but three. The submarine alone has been engaged in continuous development and improvement. In terms of performance, submarines have developed radically through the course of the twentieth century. Who would have thought in 1901 that a submarine could make a submerged circumnavigation of the world – while the development of new weapons systems has given them both flexibility and a dreadful potency. So much has changed, yet so much remains the same. Submarines are still uncomfortable and cramped places to work and the skills needed to make a successful attack are still very much the same. The submariner of the late 1990s still faced the same risks as his predecessors of ninety years ago for the sea is unforgiving and totally impartial.

As a concept the submarine has been a feature of man's imagination for centuries. Alexander the Great is said to have made a dive in such a craft in the waters off Alexandria. Yet it was not until the twentieth century, when technological developments were able to turn the concept into a reality and not until the late 1950s, with the introduction of nuclear-powered submarines, that the 'true' submarine – one that has limitless underwater endurance – has been created. The history of the submarine is littered with exaggerated claims, false starts and improbable – not to mention lethal – designs.

The Dutchman Cornelius van Drebbel is credited as being the 'Father of the Submarine'. He designed a boat in which a very timorous King James I of England was sedately rowed down the Thames in 1606. However, Drebbel's boat cannot have been anything more than a covered rowing boat, with no air supply other than what was inside the boat, and with the capacity to be ballasted down so that it ran awash if not actually submerged.

About 150 years later, David Bushnell, an American, designed a craft which has the distinction of launching the first submarine attack in history. David Bushnell had a lively and enquiring mind but was physically unprepossessing and suffered from poor health. He carried out a number of experiments to establish the destructive effects of underwater explosions and these experiments with mines (or torpedoes, as they were known in early days) led him inexorably towards the development of a craft to deliver the weapon. A small boat was out of the question, as were the clumsy divers' suits of the day. The answer lay in a submersible: a craft that could approach the target unseen. Bushnell's experiments took place at a time of rising tension in North America between those colonists who wanted self-government and the British authorities who wished to maintain the link with Britain. This lent a sense of urgency and purpose to Bushnell's work, and while the political situation deteriorated into open warfare, Bushnell and his brother Ezra worked on the craft at their farm at Saybrook, New York state.

We know very little about *Turtle*, as Bushnell's craft was named, largely because Bushnell himself destroyed the plans to prevent their falling into the hands of the British. Bushnell's own account remains the most comprehensive. In a letter dated 13 October 1787 to Thomas Jefferson, Bushnell described his submarine thus: 'The external shape of the submarine vessel bore some resemblance to two upper tortoise shells of equal size joined together. . . . the inside was capable of containing the operator and air sufficient to supply him thirty minutes.'[1]

Turtle was a one-man, hand-operated craft. David Bushnell was not strong enough to operate the craft, so his brother Ezra was the first to take the craft into the water. The 'weapon' was a 150lb charge of gunpowder, packed in a watertight package and fitted with a clockwork fuse. The charge was attached by a lanyard to an auger – a large pointed screw that could be worked from within *Turtle* so as to pierce the wooden hull of the target and secure the charge. When the charge was safely secured, *Turtle* would retire. The only instrumentation was a compass, lit by fox-fire (rotten stumpwood) and a primitive depth gauge, 18in high and 1in in diameter. The depth gauge was open to the sea at the bottom end, and depth was indicated by a floating cork.

Propulsion was by a two-bladed propeller worked by the operator's feet. If the operator was pedalling flat out, a top speed of 3 knots might have been possible but it would have been exhausting work and would have left the operator too tired to do much else. Some reports speak of a hand-cranked propeller, but there is no doubt that it would have been much easier to use feet rather than hands. *Turtle* was kept stable by means of a 200lb mass of lead ballast suspended beneath her by means of a 50ft rope. The weight of this ballast, combined with that of the operator, also sufficed to trim *Turtle* so low in the water that only the top of the hatch and the glass viewing ports were visible from the surface. The craft could be further submerged by means of a vertical screw oar and by admitting water to a small tank in the bottom of the craft which could be pumped

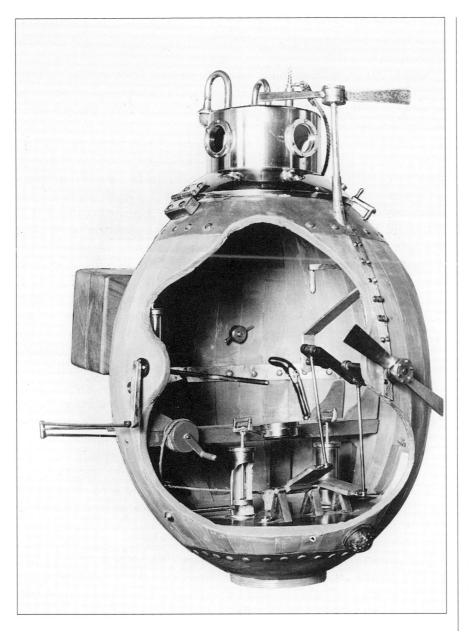

A model of Ezra Lee's Turtle, *a simple one-man craft which acted as inspiration for submarine designers and operators alike. (Author)*

out by hand if necessary. In theory it was possible to submerge the craft to a depth of several feet with a reasonable degree of accuracy and control. The operator was supplied with air by means of tubes passing through the upper hatch and fitted with shut-off valves.

By the spring of 1776 *Turtle* had been shipped down to New York where British ships were known to be anchored and where the insurgent forces possessed a secure base on the Battery (the southern tip of Manhattan Island). On 13 July 1776 the British 64-gun ship-of-the-line HMS *Eagle*, wearing the flag of Adm Lord Howe, arrived from England and moored off Staten Island. A more obvious or suitable target could not have presented itself. Just as *Turtle* was ready, misfortune struck when Ezra Bushnell fell ill and it was obvious that he would not be fit for the planned operation. Faced with pressure from Gens Putnam and

SUBMARINE ACTION

Washington to proceed with the operation, a volunteer was sought to take Ezra Bushnell's place.

It was a serious undertaking: to find a man who could learn in a matter of weeks what Ezra Bushnell had spent five years perfecting. However, one Sgt Ezra Lee stepped forward and was accepted. Even though Lee underwent arduous training in Long Island Sound, it was inconceivable that he could equal the skill of his predecessor. Nevertheless, on the night of 5/6 September 1776 Ezra Lee stepped down into *Turtle*. A rowing-boat came alongside and, with a full moon and an ebb tide, the boat made off downstream with *Turtle* in tow to where *Eagle* lay, some 5 miles distant.

It is at this point that apocryphal accounts of Lee's voyage part company from what we can deduce from known facts and experience of submarine operations. The accepted version is that Lee navigated his way downstream until he reached *Eagle*. The current was so strong that he was swept past the flagship, and he had to pedal furiously in order to regain position. Once under *Eagle*'s stern, he attempted to secure the charge to her hull but was defeated by the copper sheathing that covered the hull. With day breaking and the tide flooding, Lee gave up and made his way upstream. *En route* he was spotted and chased by a British guard boat. In an effort to distract his pursuers, Lee jettisoned the explosive charge, which is supposed to have drifted northwards and exploded at the entrance to the East River. Meanwhile, *Turtle* had been spotted by friendly forces and a boat was despatched to bring the exhausted Lee back to the Battery and safety. Subsequently, at least two attacks were made on shipping above New York in the Hudson River, but to no effect.

This is the version of the attack which every schoolboy knows and which has entered the literature of the American War of Independence. However, using our knowledge of midget submarine operations gained during the Second World War, certain facts can be established. Firstly, it was inconceivable that Lee could have made his way down to *Eagle* and positioned *Turtle* directly under her hull: it was hard enough for modern midget submariners to position their craft accurately with the aid of reliable motors, controls and an accurate compass. Lee would have had to be his own commanding officer, navigator, helmsman, look-out, propulsion and, eventually, weapons engineer. He simply had too much to do – a failing of all such one-man operated craft. It is possible that Lee managed a form of controlled collision with *Eagle*. However, *Eagle* was well guarded by Royal Marine sentries whose standing orders were that the 'All's Well' was to be given every ten minutes. Moreover, in order to keep the men awake, they were to be relieved regularly and at short intervals. It is inconceivable that *Turtle* could have made her approach and a very noisy submergence under the noses of alert guards. Furthermore, there is no reference in *Eagle*'s log to any attack by an underwater craft or, indeed, to anything out of the ordinary – including the explosion of the charge at the entrance of the East River.[2] Finally, the myth that the attack forced Lord Howe to lift his close blockade of New York is not true. The fleet remained off New York until January 1777.

So what did happen? There is no doubt that *Turtle* existed and that attacks were mounted on British ships. However, the exaggerated accounts of Lee's operations were probably the result of the insurgent forces making

4

the most of a useful propaganda opportunity. The physical effort required to operate *Turtle*, coupled with increased rate of respiration as the result of Lee's apprehension at undertaking the mission, would have seen off the 30 minutes' air supply in no time at all: Lee was probably overcome by carbon dioxide during the operation. CO_2 poisoning would account for Lee's symptoms of confusion, anxiety and generally weakened condition which historians can infer from the various accounts. In the words of one commentator, Lee would not have known 'whether it was Christmas or Marble Arch'.[3] What most probably happened was that Lee set off, then as CO_2 poisoning set in, he merely bumbled around the Narrows getting weaker and weaker, until he decided to give up and try to return. On the return journey, his compass stopped working and he had to open the hatch to see where he was going. The fresh air saved his life.

None of this detracts from the fact that Lee was a very brave man and that the whole operation was a courageous and daring effort. The legend of what happened in the Narrows on the night of 5/6 September 1776 inspired the early pioneers of the modern submarine. Today it is doubtful that one in a thousand of the office workers who crowd into their office blocks in Manhattan realise the significance of what took place beneath their office windows 300 years ago.

After *Turtle* there were innumerable attempts at building the perfect submarine. Some, like the Revd Garrett's *Resurgam*, were of doubtful

Holland No. 2, one of the first British submarines and similar to boats serving in the US Navy. This photograph shows her at Barrow on completion. (Author)

utility and were likely to be more lethal to their operators than to their opponents. Nevertheless, there was continuous demand for a craft that could operate underwater and launch attacks on enemy shipping while remaining undetected. The French came close to success with *Gymnote* in 1888 and even closer to success with *Narval* in 1899. However, the true father of the modern submarine is the Irish-American J.P. Holland, who built his first submarine in 1878. By 1900 he had built *Holland VIII* which was a great success and led to orders from both the US Navy and the Royal Navy.

The pace of development was inexorable: HM Submarine No. 105 was 120 tons and armed with one torpedo. Thirteen years later HM Submarine *E.1* was 655/796 tons and armed with four torpedo tubes and four reloads. In that period of time petrol engines had given way to diesels, submarines were fitted with wireless, they undertook long ocean passages (thus disproving the notion that they were fit for nothing more than coastal operations) and last, but by no means least, they were fitted with 'heads' which could be flushed straight into the sea – a small thing perhaps but one that improved the lot of submariners no end. By 1914 the diesel-electric submarine had reached the pinnacle of development. The British E-class and the German U31 represented the most that could be achieved with the technology that was available. Boats would get slightly larger, possess better sensors and be a good deal more comfortable, but in terms of general performance there was not much difference between a

The old and the new: Holland 3 *under way in Portsmouth Harbour with HMS* St Vincent *in the background. (Author)*

A contrast in styles. The British K.7 alongside the German U70 after the First World War. U70 sank fifty-three merchant ships totalling 137,717 GRT and 1,290 tons of warships; K.7 fired one torpedo at a U-boat – which missed. (Author)

British E-class of 1914 and a T-class of 1939 (or a German U31 type and a Type VII). Of course, there were certain exceptions to this rule, such as the British K-class submarines, but these lay outside the main-stream of submarine development.

The pace of development hid the fact that the naval authorities had very little idea of what to do with these craft. It was as if each country had submarines because every other country had submarines. The submariners themselves were confident of what they could achieve, and demonstrated their prowess in exercises, but all too often these 'successes' were dismissed by a conservative naval establishment. It took the shock of war to demonstrate the awesome potential of this new weapon. Germany began the war by using her submarines as an advance warning screen against a British attack. U-boats were secured to buoys in the Heligoland Bight and left there for a day before they were towed back to harbour. However, there were those in the German Navy who saw the U-boat as a means of countering the British blockade and starving Britain into surrender by attacking British merchant shipping. This policy, known as Unrestricted Submarine Warfare, meant the abrogation of the conventions in international law that indicated which cargoes could be sunk. Henceforth all British and Entente merchant shipping would be sunk on sight and without warning. In 1915 and again in 1916 the Germans announced such a policy but on each occasion diplomatic pressure (largely from the United States) brought about the termination of the campaign. Finally, in 1917, with the country facing starvation, the German government embarked on their third and final campaign of Unrestricted

Defeat, 1918. Austrian U-boats lie abandoned at Pola awaiting division among the Allied powers. (Kriegsarchiv, Vienna)

Submarine Warfare. But by this time sufficient counter-measures (notably the convoy system) were in place to contain the offensive. The other effect of the German 'stop-go' policy was that they never had enough of the type of submarine they required. Construction programmes generally lag behind political decisions by nine months. Thus coastal submarines of the UB and UC type for *Handelskrieg* were ordered but by the time they had been built the policy had changed and submarines for long range operations in accordance with international law were required.

British submarines did not have the same opportunities as their opponents across the North Sea. With the imposition of the blockade, German maritime trade disappeared from the high seas. Instead British submarines were sent into waters that could well be described as the 'enemy's back yard': the Baltic, the Sea of Marmara and the Adriatic. There they disrupted German, Turkish and Austrian coastal traffic. However, the inessential nature of this traffic meant that the British submarines could never have the dramatic effect that the U-boats did. In the last two years of the war, British submarines were increasingly employed on anti-submarine operations – a trend nearer that of our own time. The French were limited by the same constraints as the British.

Opposite: *Victory, 1918. A German UC-class minelayer wears British colours over the German ensign as she is taken over by a British crew. (Author)*

Linienschiffsleutnant Egon Lerch salutes from the conning tower of his U12 *after torpedoing the French battleship* Jean Bart *on 21 December 1914. Although the battleship returned to port, fear of further attacks caused the French to keep their capital ships out of the Adriatic. This was a striking demonstration of the influence exerted by the submarine. (Kriegsarchiv, Vienna)*

However, French submariners won a reputation for desperate bravery by entering the heavily defended ports of Cattaro and Pola. That they did so in unreliable steam-driven submarines makes the feat more remarkable. Austrian submarines effectively barred the French from the Adriatic until Italy entered the war in 1915. Although Austrian submariners were extremely proficient, their operations have been unjustly overshadowed by those of their German allies. Lastly, Russian submarine operations were extremely ineffective. Although the evidence is that their submarines were well built, the personnel were poorly trained and motivated.

After the First World War all the victorious powers reduced their submarine fleets in line with the various disarmament treaties and as a result of the need for financial economy. Germany, of course, did not enter this equation since she was banned from operating submarines under the terms of the Versailles treaty. Austria-Hungary, the other submarine power on the losing side, no longer existed and thus was also out of the picture.

During the interwar period all powers pursued a two-track policy as regards their submarine programmes. Diplomatically, Britain (the chief victim of submarine warfare in the First World War) and a number of other countries sought the abolition of the submarine as a weapon of war. This was an ideal which lay beyond the bounds of achievement, and most submarine construction programmes implemented by the main naval powers took place against this background. Nevertheless, this period saw some interesting new trends. Britain now regarded the submarine as her main strike weapon in the Far East against Japan and thus British

The Austrian U5 *entering Cattaro harbour in August 1915. Note the camouflaged conning tower and, most unusually, an unguarded forward hatch. (Kriegsarchiv, Vienna)*

submarines, like their American counterparts, were built for long-range offensive operations and were at the forefront of British naval planning in the Far East. It was perhaps unfortunate that this concept hinged on Britain retaining the use of the bases at Hong Kong and Singapore. Once Germany was allowed back into the submarine community after the 1935 Anglo-German Naval Agreement, German planners sought to begin where they had left off in 1918, using the U-boat against British trade. However, the leadership of the *Kriegsmarine* was totally focused on the battleship as the *ne plus ultra* of naval warfare and thus by the time the Second World War broke out in 1939, there were not sufficient submarines to implement a trade war on the scale demanded by the U-boat men. The Soviet Union also built a large submarine fleet during this period but tight political control hindered the development of sound operational or tactical concepts.

In 1939 the world was once again at war and submarine operations followed a pattern very similar to that of the First World War. German U-boats sought to starve Britain into surrender by attacking Britain's maritime trade, but this time they had the advantage of possessing bases on the French Atlantic coast. Thus it was not so easy to blockade the U-boats using minefields and net barrages as it had been in the First World War. The Germans waged a tightly controlled offensive against British

British E- and H-class submarines – all veterans of the war in the Mediterranean – lie at 'Rotten Row' in Grand Harbour, Valletta. Submarine fleets shrank drastically with the end of hostilities in 1918. (Author)

The Type VIIC U-boat was the workhorse of the German Second World War submarine fleet. This particular boat is U570 which is shown under British colours as HMS Graph following her capture in August 1941. (Author)

The launch of USS Cisco *on 24 December 1942 at Portsmouth Navy Yard, New Hampshire. American industry proved more than adequate at meeting the demands of the US submarine fleet and by 1944 the US Navy had more submarines than it could crew. (US Navy)*

shipping in which U-boats operated largely on the surface in night attacks against convoys. The system depended largely on tight control by shore headquarters and reliable and secure communications.

Initially the campaign went in favour of the U-boats but by mid-1943 the Allies had developed sufficient experience and counter-measures to force the U-boats to withdraw from the Atlantic. The German response was to develop faster and deeper-diving submarines, together with new weapons such as the homing torpedo. Fortunately, from an Anglo-American perspective, these submarines arrived too late to influence the war and counter-measures were swiftly developed to beat the new weapons. However, the British official historian was forced to admit that the vast anti-submarine effort deployed by Britain, Canada and the United States only succeeded in containing the U-boats and not defeating them.

In the Pacific the submarines of the US Navy launched a devastatingly effective campaign against Japanese maritime trade which succeeded in starving Japan into surrender. The Japanese neglected the whole field of anti-submarine warfare (ASW) and paid the price accordingly. Meanwhile, the large and technologically impressive Japanese submarine fleet remained tied to the apron strings of the battle fleet in a reconnaissance role and, as the war progressed, acted as a supply service to Japan's beleaguered island possessions. The immense American supply line, running from the Pacific coast to Hawaii and then on to the various fronts in the Pacific, was ignored by Japanese submariners and their failure to exploit this area of operations must rank as one of the greatest missed opportunities in naval history.

British submarines fought a very different war. Germany and Italy were not maritime powers and therefore had no vital maritime trade that could be attacked. Instead, British submariners carried out their operations in shallow waters along the enemy coast. These waters were often mined and lay close to concentrations of enemy ASW forces. In home waters British submarines were kept on a tight leash and their deployment was influenced almost totally by the need to guard against a breakout by the German battleship *Tirpitz*. It was not until the last year of the war that they were free to attack German coastal traffic off Norway. In the Far East the magnificent British submarine squadron had been withdrawn to the Mediterranean by the time of the Japanese declaration of war and it was not until early 1944 that British submarines reappeared in this theatre in any numbers. Operating from bases in Ceylon and Australia, British submarines were confined to the Malacca Straits and the waters off Malaya.

It was, however, in the Mediterranean that British submarines were at their most effective. Operating from Malta, in very shallow waters which were close to enemy bases and very heavily mined, they played a significant part in disrupting Axis convoys taking supplies to North Africa.

In the postwar period submarine development took another radical step forward. The Germans had invested considerable resources in developing a submarine with enhanced qualities of underwater endurance. These submarines, the Type XXI and Type XXIII, were intended to win back the initiative in the Battle of the Atlantic but hostilities ended before they could be deployed operationally. Instead the British, Americans and Russians inherited the benefits of this technology and were not slow in applying it. In particular the rapid growth of the Russian submarine fleet following the breakdown of the wartime alliance acted as a stimulus to western development. Russian submarines posed the same threat to western maritime trade as had the German U-boats, yet the west could not afford to maintain and deploy substantial surface ASW groups in the face of this threat. The submarine was seen as the cheapest and most effective way of countering the Russian submarines. Yet the wartime diesel-electric boats, festooned with guns, were extremely unsuited to this task. They needed to be faster, quieter and equipped with sonars capable of detecting and tracking a target while both parties were submerged. The effects of this requirement were visible almost immediately. Guns were removed and superstructures cleaned up to reduce water resistance and noise, and the submarines were equipped with the Schnorkel, an underwater breathing tube that allowed them to use their diesels while dived.

The Royal Navy converted several T-class submarines into 'Fast Battery Drive' boats by lengthening the hull for the insertion of an extra battery and enclosing the pressure hull in a streamlined casing with all masts contained within a 'sail' or 'fin'. Several more T- and A-class submarines received a less extensive conversion. This gave the boats much greater

USS Catfish *(SS339), a GUPPY II conversion photographed in 1962. These were war-built submarines with all unnecessary upper deck fittings either removed or enclosed to give a totally streamlined form. The conversion had the effect of making them faster and quieter.* Catfish *was subsequently sold to Argentina, renamed* Santa Fe, *and sunk by the British in April 1982 during the Falklands War. (US Navy)*

underwater speed and endurance. In the US Navy the Greater Underwater Propulsive Power (GUPPY) programme achieved the same objectives. But while the programmes provided an interim solution, they did not fulfil the designers' requirements for a true submarine. The British T-conversions and American GUPPYs still had to surface or schnorkel to charge their batteries. Each time they did so, they were exposed to detection. What the submariners and the designers wanted was a true submarine – one that did not have to surface or schnorkel to charge the battery. Such a submarine would require a powerplant that was independent of outside air.

The Royal Navy initially felt that the hydrogen peroxide and the Walther engine would provide the solution. Experiments with a captured German U-boat, HMS *Meteorite*, and two specially built trials submarines, HMS *Excalibur* and HMS *Explorer*, proved otherwise. While the submarines displayed increased performance, the highly volatile nature of hydrogen peroxide made it an unsuitable fuel source for a submarine. Not for nothing were the two boats known to their crews as *Exploder* and *Excruciator*. The Americans on the other hand had opted for nuclear power, building on the large amount of research work carried out in connection with the Manhattan Project. The first nuclear-powered submarine, USS *Nautilus*, was launched in 1954 and, although in terms of hull form she resembled an improved German Type XXI, with her top underwater speed of 23 knots she was able to outrun any surface opposition. The adoption in 1956 of the 'teardrop'-shaped hull form in the USS *Albacore* was of immense importance. The teardrop-shaped hull reduced water resistance and water-flow noise around the hull and increased speed. This hull form has now become standard for submarines all over the world.

Nautilus was followed by a series of bigger, faster, deeper diving and more powerfully armed 'fleet boats', culminating in the 688 Los Angeles class which today forms the backbone of the USN's SSN fleet. Britain's first nuclear submarine, HMS *Dreadnought*, was completed in 1963. Today the Trafalgar-class SSNs are among the most sophisticated in the world. However, the high cost of the SSN programme means that in America the successor class to the 688s, SSN21 USS *Seawolf*, is limited to three units and has been the subject of fierce public debate. In Britain the replacement for the Trafalgar class will be no more than a repeat

USS Helena, *one of the 688-class SSNs which currently make up the bulk of the American submarine fleet. Armed with a mixed load of cruise missiles, torpedoes and/or mines, they constitute formidable fighting vessels. However, some believe that the design is too large for inshore or under-ice operations.* (US Navy)

HMS Finwhale, *a British Porpoise-class SSK, shows off her paces on the surface in 1973. The Porpoise-and subsequent Oberon-class SSKs were extremely successful and variants have been sold to Australia, Brazil, Canada and Chile. (Royal Navy)*

construction. Indeed the cost of nuclear-powered submarines means that Britain and the USA now have all-nuclear submarine fleets, diesel-electric boats having been deleted on grounds of cost. The USSR's first SSNs, the November class, were extremely noisy and poorly regarded. Since then the Russians have gradually closed the quality gap and the new Severodvinsk-class SSN is expected to be the equal of anything in the west.

One further advantage of the introduction of nuclear power in submarines was that it facilitated the ballistic missile submarine concept. Submarines are the ideal carriers for a nation's ballistic nuclear missiles: they are almost immune from a pre-emptive first strike, which could easily destroy a similar land-based facility such as airfield or silo, yet have the capacity to be instantly available should the need arise.

The use of nuclear power has also opened up a new area for submarine operations: the seas under the ice-cap around the North Pole. In the 1950s and '60s a number of SSKs made daring forays under the ice but these voyages were limited since the boat had to be back out in open water in order to schnorkel or surface for the all-important battery charge. Nuclear-powered boats suffered no such constraints, as was graphically shown by the submerged transit of the North Pole by USS *Nautilus* in 1959. Under-ice operations by the nuclear-powered submarines of all

HMS Churchill, *a second-generation British SSN. (MoD)*

navies, including visits to the North Pole, are now almost routine. The Russians in particular have paid a good deal of attention to this area and a senior Russian naval officer was recently quoted as saying, 'He who controls the seas under the ice can control the world.'

There are eight roles which can now be performed by either an SSN or SSK.

• Co ordinated High Intensity Strike: using cruise missiles such as the Tomahawk to influence a land battle by posing a serious threat in the period prior to hostilities and after hostilities commence. The submarine can operate from a position of safety and yet destroy targets that might be otherwise invulnerable.
• Anti-Submarine and Surface Unit Warfare: this is arguably the most important role of the submarine.
• Surveillance: the submarine has the capability to approach close to opposition forces and monitor their operations and movements using visual and electronic means. Surveillance also includes the important role of beach reconnaissance. Using modern video technology a

submarine can make a significant contribution to the intelligence collection effort prior to an amphibious landing.

• Regional Sea Denial: the ability to project a submarine forward into the opposition's waters and thus reduce his flexibility of operation. This role ties in closely with;

• Covert Power Projection: the 1982 Falklands War provided a classic demonstration of this aspect of submarine operations. British SSNs arrived off the Falklands before the main body of the task force and exercised considerable influence over Argentine plans.

• Offensive Inshore Operations: blockading an enemy harbour or protecting the approaches to an area selected by the opposition for an amphibious landing.

• Covert Operations: the submarine remains the ideal instrument for the insertion and recovery of special forces.

• Support of Amphibious Landings: submarines can be readily available for a flexible response against any surface or subsurface threat.

The twentieth century has been truly that of the submarine. The transformation from the small, unreliable and unproven vessels of 1905 to the nuclear-powered leviathans of the new millennium is remarkable, and the appetite of the world's navies for more submarines does not seem to have diminished. Exports of diesel-electric boats to all parts of the world from European countries, either new construction or old boats being sold on, are the only facet of the arms trade that has proved immune to the various postwar recessions. This book is the story of the submarine in the twentieth century.

NOTES

1   Transactions of the American Philosophical Society for Promoting Useful Knowledge: *General Principles and Construction of a Submarine Vessel*, communicated by D. Bushnell of Conn. (the Inventor), in a Letter of October 1787, to Thomas Jefferson, then Minister Plenipotentiary of the United States at Paris; Vol. IV, 1799, p. 303.

2   PRO: Journal of the Proceedings of His Majesty's Ship *Eagle*, by Capt Henry Duncan, between 10 February 1776 and 28 February 1777.

3   Compton-Hall, Richard: *Monsters and Midgets* (Blandford Books, 1985) p. 93.

SUBMARINE
ESSENTIALS

All submarines of whatever vintage and state of development have to meet the following criteria: they must be able to manoeuvre freely in three dimensions over a considerable period of time, to be able to see the target while submerged and to be able to launch a reliable and effective weapon. None of these requirements is as easy to fulfil as it seems. To begin with, any submarine must have a hull that is sufficiently strong to withstand the pressure of the sea at depth.

The main structure of a submarine, known as the pressure hull, must resist the water pressure at the greatest depth the submarine may reach. The term *Diving Depth* covers a multitude of meanings – some deliberately configured to deceive a potential enemy – but the following definitions may be useful. *Operational Depth* refers to the maximum depth to which the boat can be taken in the course of normal operations. Operational depth always incorporates a considerable safety margin to allow for the occasional deep dive. A British T-class submarine of Second World War vintage had an operational depth of 300 ft, but on 23 April 1940 *Tetrarch* was forced down to 400 ft plus during a determined German depth-charge attack. *Test Depth* is reached during a new submarine's trials when the boat is taken to the maximum operational depth. Some COs take the boat 10 per cent deeper in order give their ship's company confidence in the strength of their boat. *Collapse Depth* is the depth at which it is estimated that water pressure would cause complete failure of the hull structure.

Within or around the pressure hull ballast tanks can be arranged in three different ways: single-hull, double-hull and saddle-tank. In a single-hull arrangement, all main and compensating tanks together with fuel tanks are contained within the pressure hull. Such was the construction of many early submarines. In single-hull submarines all internal tanks which are flooded from the sea must be able to withstand full sea pressure at depth – for obvious reasons – or a valve must be provided to close off the tank from the sea as soon as the boat has dived. The first solution is difficult to achieve as ballast tanks, by the nature of their construction, have large flat surfaces that are ill-suited to withstand pressure. The second is equally impractical since the fitting of such a valve means that the tanks cannot be blown at depth. Single-hull designs are therefore only really suitable for a limited diving depth. A further disadvantage to the

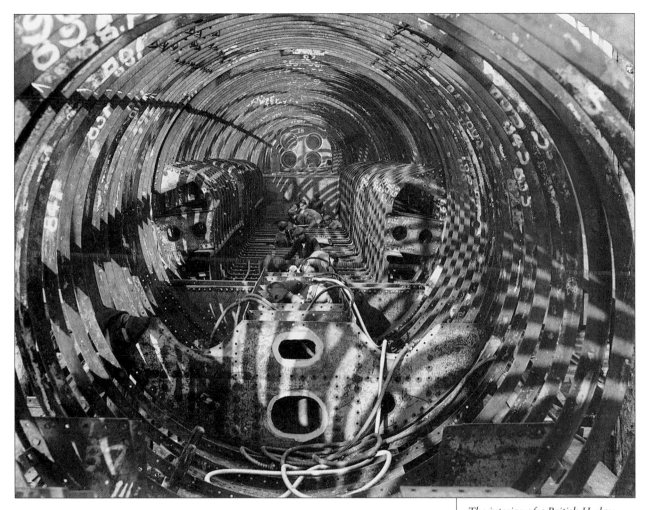

single-hull design is that having all the tanks within the pressure hull makes for a very cramped interior.

In a double-hull design all tanks are contained in a space between the inner pressure hull and an outer skin. This design radically increases weight but the increased buoyancy improves surface sea-keeping. Double-hull designs were extensively used by the French and Italian navies before the First World War and by the German Navy during both world wars. The saddle-tank design is an intermediate version in which some tanks, notably the compensating and auxiliary tanks, are fitted internally but the main ballast tanks (MBTs) are fitted as 'blisters' along the port and starboard sides of the casing. This arrangements permits the tanks to serve a dual purpose, both as ballast tanks and as additional fuel tanks. During the Second World War the British T-class submarines serving in the Far East had their range increased to 11,000nm as a result of being able to carry fuel in external ballast tanks.

Modern British and American designers favour single-hull submarine designs. The American SSN688 is a particularly unusual example with all ballast tanks concentrated at bow and stern between the pressure hull and the external skin of the submarine. By contrast Russian designers seem to

The interior of a British H-class submarine seen under construction in 1917. The view is looking forward from aft with the apertures for the four 21in torpedo tubes clearly visible at the far end. The workmen in the midships portion of the hull are working on the battery section. Because of the single-hull design, the main and auxiliary ballast tanks were placed under and beside the battery tanks, where their tops extended well above the battery tanks as can be seen in the photograph. (Author)

The launch of the British submarine Unity *in 1939. This photograph gives an excellent view of a single-hull design where all ballast and fuel tanks are contained within the pressure hull. (Author)*

favour the double-hull design. Russian submarines spend a good deal of time operating in ice fields, so a 'second skin' gives additional protection in case of a rupture to the hull. A double-hull also offers some protection from a torpedo hit.

What shape should a submarine's hull be to withstand the pressures of the sea? The ideal shape to withstand pressure is a sphere, followed closely by a cylinder. Bushnell's *Turtle* was nearly a sphere but there was no practical way in which a spherical shape could accommodate all the machinery needed by a submarine – even the earliest ones. The submarines built by John Holland adopted a spindle hull shape with a wide 'middle', tapering sharply at the bow and stern. Although Holland was probably unaware of it, this shape was ideal for the best underwater performance. On the surface, however, it was different matter: the *Holland* boats in the Royal and US Navies handled appallingly.

As the endurance and capabilities of early submarines increased, it became clear that they would spend a lengthy time on the surface travelling to and from their patrol areas, diving only when in the immediate operational area or when attacking (or being attacked). Accordingly, they were given a more 'ship-shape' structure; built around

A British E-class submarine in the floating dock at Harwich during the First World War. The photograph shows the saddle-tank arrangement on the boat's starboard side. (Author)

The British submarine HMS Tapir being broken up at Faslane in 1967. The photograph was taken from over No. 3 battery compartment looking forward into the remains of the accommodation space. The circular line of the pressure hull can be seen together with the saddle tanks on each side of the submarine. The top of the pressure hull has been lifted to allow the removal of large items of equipment. (Dr Ian Buxton)

The Italian submarine Jantina *was typical of Italian inter-war submarine designs. Note the large conning tower structure which often contained additional galley facilities and washrooms.*

the pressure hull, this was called the casing. This structure enhanced surface performance and provided a convenient platform for the ship's company to work on when entering or leaving harbour. The space between the casing and the pressure hull was free-flooding: when the submarine dived water flooded through holes cut into the side of the casing. This space also provided a convenient area to store such items as hawsers, bollards and other miscellaneous equipment. (See the chapter on blockade-breaking for even more outlandish uses of the casing space.)

The requirement for submarines to have almost identical surface performance qualities to those of a destroyer resulted in a neglect of diving performance. By 1943 nearly all submarines were configured for surface operations and were built with large conning towers bristling with AA guns, periscopes, WT masts and radar masts. Guns, and a number of other items of equipment, were mounted on the casing fore and aft of the conning tower. While very useful for the type of warfare that submarines were engaged in, these features did not enhance underwater performance. The large conning towers and the large number of free-flooding holes cut into the casing increased underwater drag and also made a considerable amount of noise.

During the Second World War the Japanese and the German Navy began to see the development of a submarine with a high underwater speed as one way of regaining the initiative in the submarine war. A high underwater speed would enable the submarine to outrun pursuing escorts which were not usually capable of sustaining high speeds. In fact the

Japanese had pioneered this area of submarine development as early as August 1938 with the construction of the experimental *Vessel No. 71*.[1] This submarine was designed for a surface speed of 18 knots and a dived speed of 25 knots. Her hull, shorn of all protuberances such as guns, periscope standards, stanchions etc., is uncannily like the contemporary SSKs. She was not a success and was broken up in 1940. The Japanese next designed the three ST-type submarines and the smaller STS which followed similar principles. Across the other side of the world the Germans introduced the very similar Type XXI and Type XXIII U-boats while the Royal Navy converted a number of S-class submarines for ASW training.

All these submarines had the following features in common. Their hulls were absolutely clean, with nothing in the way of obstructions. All fittings on the casing were recessed and, in the case of the Japanese boats, the deck guns could be retracted into the casing. The forest of masts and periscopes which had distinguished wartime submarines was now enclosed within a structure which came to be known as the sail or fin. These two factors reduced underwater resistance and noise and created a faster submarine. More powerful electric motors supplied by high capacity batteries provided enhanced underwater speed and endurance.

After the Second World War nearly all submarines shed their guns and other impedimenta extremely quickly and were either totally converted to a streamlined appearance and given extra motors and high capacity batteries (the British T-conversions and the US Navy's GUPPY conversions) or had a simple streamlined casing built over the existing hull

The S-class submarine HMS Seraph *shown in her 'streamlined' form minus guns and all unnecessary deck fittings.* Seraph *and a number of others were thus converted to provide realistic 'loyal opposition' to British ASW forces preparing to meet the German Type XXI and Type XXIII U-boats at sea. (Author)*

Going down! HMS Acheron *dives in 1970. (MoD)*

as in the British A-class. These changes gave radical improvements in performance. Trials between a streamlined British T-class submarine, HMS *Tireless*, and her unconverted sister, HMS *Tudor*, showed that not only could *Tireless* go faster (9.5 knots at periscope depth as opposed to 8.1 knots) but could do so while making less noise! Cavitation occurred at 2.3 knots with *Tudor* but at 3.23 knots with *Tireless*. These high underwater speeds seemed to restore some advantage to the submarine in their battle with surface vessels. Capt John Coote, who commanded the converted British HMS *Totem*, recalled an exercise in the Mediterranean in 1954: 'It was money for old rope. We'd just pop under [the] layer, wind on 18 knots and then we'd clear a mile in four minutes. Follow that with ten minutes silent running under the layer at 12 knots and we would be 3 miles from the escort.'[2]

However, one problem associated with these new high underwater speeds was that the submarine tended to pitch up or down. The problem was first noticed with captured Japanese and German craft and then in British and American postwar conversions. Above speeds of 8 knots small variations in pitch angle would grow to the point when control was lost. This meant that high underwater speeds could not be fully exploited

HMS Dreadnought, *Britain's first nuclear-powered submarine, at speed going down the Clyde in the mid-1970s.*

unless the submarines could be steered freely in the vertical as well as in the horizontal plane. Britain and America reached the solution to this problem almost simultaneously. The British went back to J.P. Holland's spindle shape hull while the Americans tested out a number of hull forms based on airship design. The result was the USS *Albacore* which was commissioned in 1956. She had a hull shaped like a teardrop (very like that of Holland's earliest boats) and could *sustain* [author's italics] without risk the high underwater speeds that were theoretically possible. *Albacore*'s shape has now been adopted for all submarines built throughout the world.

All submarines have to obey Archimedes' principle that weight equals buoyancy. Ballast tanks provide the means by which a submarine can dive and surface again. When empty of water and filled with air the tanks provide a reserve of buoyancy on the surface. When filled with water they allow the submarine to submerge with neutral buoyancy. However, buoyancy is a variable factor and can alter according to conditions of temperature, operating depth, specific gravity of sea-water and the changing weight of the boat as fuel, food and ammunition are used up. Compensating tanks are therefore required so that the weight of the boat can be finely adjusted to local conditions and so that the position of the centre of gravity does not move.

Submerged stability requires that a submarine's centre of gravity (G) lies below the centre of buoyancy (B). A value of 24 cm for BG was generally thought to be adequate. In order to surface, water was blown out of the ballast tanks using high pressure (HP) air. However, HP air was a precious resource, not to be wasted, and was thus only used to bring the boat to the surface where she lay awash. Low pressure (LP) air was then supplied from a compressor to bring the boat to full buoyancy. Venting the diesel exhaust into the ballast tanks was another option.

A submarine must be able to hold a steady course, to change direction rapidly and return to a steady path in both the horizontal and vertical planes. In particular the submarine must be able to maintain a precise depth when using the periscope despite wave motion. The solution lies in the provision of 'wings' (similar to an aeroplane's elevators) called hydroplanes (one pair close to the bow and the other close to the stern). Horizontal control is provided by a conventional rudder. Bow hydroplanes can be retracted into or folded against the hull for easier manoeuvring on the surface.

NOTES

1   The name was intended to deceive the curious about her true status.
2   Capt John Coote to author, 17 May 1989.

Opposite: 'Bacchus', the mascot of the Free French submarine Rubis, *ensures all is well in the control-room while the submarine is on patrol. On his right are the hydroplane operators while the OOW is looking up the ladder into the* Kiosque. *Bacchus went on over thirty-five wartime patrols and more than earned his retirement to France after the war. (ECPA)*

PROPULSION

The principal problem of submarine development that faced designers until late in the nineteenth century was how to provide a reliable powerplant which could provide sufficient performance both above and below the water. The immediate solution was the electric motor supplied from a storage battery. However, submarines powered solely by an electric motor possessed very limited endurance and would be suitable for little more than harbour defence. Using the electric motor, a submarine can only achieve maximum underwater speed for a couple of hours. Hence the need for a fossil-fuelled engine which could be used for surface running and which would fulfil the secondary role of recharging the battery via a generator. This combination was first provided by John Holland who used a petrol engine and a battery-driven electric motor.

Until the development of diesel the petrol engine was the principal means of surface propulsion. The engine was not without its drawbacks. The fumes were extremely hazardous both in their low flash-point and in the intoxicating effect they had on those operating the boat. The latter was sometimes considered an advantage by those members of the lower deck who deliberately inhaled the fumes and claimed to enjoy the results despite the inevitable hangover the next morning. But there were occasions when the entire ship's company became irresponsible owing to inhalation of petrol fumes and that was considered anything but funny. The German Navy never used the petrol engine but instead opted for the Korting paraffin engine. These engines were extremely heavy on fuel consumption and moreover belched out great clouds of white exhaust smoke. This was almost certainly why the German *U6* was spotted, attacked and then sunk by the British *E.16* on 15 September 1916.

Steam-power was another option. France used steam-power more than any other nation in submarines built before the First World War. It was used in the eighteen submarines of the Pluvoise class as well as in a number of individual boats. Interestingly enough the 700ihp reciprocating engines of the Pluvoise class gave a top speed of 12 knots in favourable sea conditions and with a clean hull. This speed was only half a knot faster than the 600bhp Sauter-Harle diesels fitted in the earlier Emeraude class. The half-knot was obtained at a cost of greatly increased submerging time, resulting from the need to close more than twenty valves and hull openings. It was also found that the heat of the exhaust gases caused the 'funnels' to expand, so that they would not retract snugly into their housings when necessary. Wartime experience quickly showed the disadvantages of this mode of propulsion. In 1914 the French submarine

Early use of steam-power in submarines. HMS Swordfish *forges her way down the Clyde in 1916. Britain and France made sustained efforts to make steam-power 'work' in submarines but it would not become a practicable means of propulsion until the nuclear age. (Author)*

Archimede was on patrol in the Heligoland Bight when one of her 'funnels' was damaged in a heavy sea. It was knocked off its guides and could not be lowered. In an instant *Archimede* was transformed from a submarine into a surface vessel of doubtful utility. However, it is worth noting that the Pluvoise class referred to earlier gave sterling service in the Adriatic.

The Royal Navy was drawn towards steam propulsion for submarines in the search for a higher surface speed. If submarines were to work with the fleet then they had to operate at the speed of the fleet and the diesels then in service simply did not generate enough power. In vain the British tried fitting J-class submarines with three diesel engines. These large double-hull boats had two engine rooms (machinery space occupied 36 per cent of the hull length) but could only attain a top speed of 17 knots – not fast enough to work with the fleet. Following the successful installation of a steam powerplant in the experimental submarine *Swordfish*, steam was selected for the large K-class fleet submarines. These boats were fitted with two-shaft Brown-Curtis or Parsons geared steam turbines with Yarrow boilers, and 10,500shp gave a top speed of 24 knots – which was exactly what the Admiralty wanted.

Sufficient ink has been expended on the K-class elsewhere but suffice to say that they were probably the most hazardous submarines ever sent to sea – and that statement includes some of the more notorious Russian designs. Eight suffered disasters and there were sixteen major accidents

The French steam-powered submarine Cugnot *in 1909. Despite problems with the steam propulsion plant,* Cugnot *and her sisters were highly successful in the Adriatic. (Marius Bar)*

The British steam-powered K.22 preparing to dive. Over forty valves had to be secured before diving – the process took over five minutes as opposed to an H-class submarine's thirty seconds! (Author)

and an untold number of minor incidents. The danger came from having too many hull openings for air intakes and exhausts for funnel gases, some of which were as large as 1m in diameter. Shutting down the steam plant and securing the boat for diving took nearly four minutes – a tremendously long time considering that a UB- or UC-class submarine could be under in less than 30 seconds.

It was the diesel engine that came to the submarine designers' rescue as the ideal surface powerplant. The diesel was adopted by the Royal Navy for the D-class in 1907, by the US Navy in 1903 and by the German *Reichsmarine* in 1912. Since then it provided the principal means of surface propulsion for the submarine until the introduction of nuclear power. Early submarine diesel-electric systems were direct drive, in other words the diesels were directly connected to the propeller shafts and had to be de-clutched before the electric motors could be engaged. When charging the battery only one diesel could be used for propulsion while the other charged the battery. However, in the large American fleet submarines of the Second World War the diesels were coupled to generators, always running the boat on the electric regardless of whether she was submerged or not. The same system was used in the British U-class. This was cumbersome but had the effect of enabling the submarine to maintain a high surface speed (particularly when moving between patrol locations or in pursuit of a target) while still being able to put a full charge into the battery.

In the postwar period the diesel has retained an important role in submarine propulsion; even nuclear-powered submarines carry a diesel generator in case of a reactor failure. However, the way in which the diesel engine has been used has changed significantly. Previously, diesel direct drive was useful when making long surface runs. However, in the postwar period the emphasis was placed on submerged performance. Moreover, as sonars developed in sensitivity it became apparent that diesels produced low frequency noise which travelled a great distance underwater, thus making the boat an easy target for ASW systems. Nowadays, modern boats use smaller, faster running diesel (1000rpm as opposed to 440rpm) to run generators which produce electric power for both motors and batteries. These diesels produce a much higher charging rate and can restore power to a battery in something under 20 minutes! Naturally, using the diesel for short periods reduces quite radically the boat's chances of being detected. The end result is a faster and quieter submarine. A British wartime A-class submarine could manage 16.5 knots on the surface but a maximum of only 8 knots dived, and that for little more than hour. The *Upholder* class of the 1990s is capable of 12 knots on the surface but (aided by a teardrop hull) can achieve 20 knots underwater using the battery.

The introduction of the schnorkel or underwater breathing tube was the most significant development in submarine propulsion up to the introduction of nuclear power, and allowed a submarine to run her diesel engines while submerged. It is claimed that the schnorkel was developed by the Dutch before the Second World War. However, two British submarines of the First World War period, *C.3* and *E.35*, were fitted with home-made 'underwater breathing tubes' designed by their commanding

The interior of a British Oberon-class SSK's engine-room looking forward. (VSEL)

officers. These were fairly hair-raising systems from a safety aspect but their appearance showed that submariners were thinking about the need to use their diesel engines while submerged. In 1940 the Dutch system was discarded by the British and Germans, both of whom acquired complete schnorkel systems following the German occupation of Holland in 1940. However, it was the Germans who first sent the system to sea in 1944 (*U264* was the first *frontboot* to be fitted with the device) in an attempt to win back the initiative in the Battle of the Atlantic. Dönitz hoped that the new system would offer his U-boats some immunity from Allied aircraft. It

The Dutch submarine O.21 demonstrating the schnorkel, or underwater breathing tube, at Flushing before the Second World War. From right to left the submarine's masts are a raised periscope, the schnorkel induction mast and the schnorkel exhaust mast. O.21 brought her schnorkel to England following the invasion of Holland, where it was dismissed as unworkable by the Royal Navy. (Royal Netherlands Navy)

would not be an overstatement to claim that the introduction of the schnorkel system gave a new lease of life to the diesel-electric submarine which would otherwise have been discarded years ago, overwhelmed by maritime patrol aircraft (MPA) with increasingly sophisticated detection devices. Even SSNs/SSBNs are fitted with a schnorkel mast for use when running the emergency diesel.

The working of the system is simple. A retractable mast is fitted with two pipes: one allows air for both crew and diesels, to be drawn into the submarine while at periscope depth. The second allows the diesel exhaust to be expelled from the submarine. Early systems consisted of a combined induction and exhaust mast of a fixed length which lay flat along the casing when not in use. After the mast had been raised it first had to be drained down into an auxiliary ballast tank before the engines could be started. Once the engines were running the exhaust pressure cleared the water from the exhaust mast system. The top of the induction mast was fitted with a float mechanism which stayed shut if the top of the mast dipped beneath the water but dropped open in air to uncover the top of the induction tube. Heavy seas or the failure of watchkeepers to keep an accurate depth could result in the top of the mast dipping beneath the surface and the valve slamming shut. Air for the diesels would then be drawn from within the boat, creating vacuum levels that represented a

The conning tower of the British T-class submarine Taciturn *photographed after the Second World War showing her schnorkel mast in the raised position. (Author)*

Opposite: *The schnorkel induction mast of a British submarine lying flat on the after casing in the 'down' position. (Author)*

dangerous lack of oxygen for the crew. When the top of the mast reappeared above the surface, the sudden rush of air into the vacuum was intensely painful. In February 1945 the crew of *U953* (KL Herbert Werner) endured an extremely unpleasant patrol as the result of the *Leitender Ingenieur*'s inability to control the boat while schnorkelling: 'Each time the Chief disturbed the buoyancy, it sent the whole company through new tortures. The vacuum it created set the men twisting and vomiting in agony.'[1]

Other disadvantages of the early systems included the fact that the diameter of the induction mast was so narrow that engine speeds had to be carefully regulated to avoid sucking air out of the submarine and creating a vacuum. In modern systems both masts are periscopic, so their height can be adjusted to meet sea conditions. The induction and exhaust are also in separate masts, thus allowing a greater volume of air to be drawn into the boat.

Despite its theoretical advantages, the schnorkel did not prove a universal panacea for the diesel submarine's problems. The raised mast increases the boat's vulnerability to detection. The mast itself creates a large wake on the surface which is increased by the 'gefuffle' from the exhaust. The mast can also be tracked by radar. (Not long after the Germans sent *U264* to sea with the first schnorkel, British ASW forces were training on attacking a schnorkelling submarine fitted to HMS *Vulpine*.) Modern ASW devices include infra-red sensors to detect the tell-tale exhaust gases. Lastly, increased use of the diesels makes the submarine more noisy and thus easier to detect, while at the same time reducing the effectiveness of her own passive sensors. These disadvantages can only be minimised, not avoided. British Oberon SSKs have developed the technique of 'gulping' – extending the induction and exhaust masts for short periods only, thus creating an intermittent radar target.

Schnorkelling also poses serious 'health and safety' concerns. While running submerged with the schnorkel system in use, the boat is effectively proceeding with two large holes in her hull open to the sea: a situation that is anathema to most submariners. Engine-room staff must be constantly alert and ready to shut down the system in case the induction mast breaks or is damaged in a collision, and the boat starts to flood. Barometers, marked with green, orange and red sectors indicating human tolerance for vacuum conditions, must be fitted in both control-room and engine-room. In addition, care must be taken not to schnorkel with the wind right astern – under these conditions the exhaust gases will be drawn back down the induction mast into the boat. However, the most dangerous use of schnorkelling is charging the submarine's batteries while dived because the process produces hydrogen. When the boat is on the surface, venting this gas is not a significant problem as a draught of air is being drawn into the boat down the conning tower. However, when dived hydrogen can build up to dangerous levels.

Such a build-up of hydrogen was almost certainly responsible for the loss of the USS *Cochino* (SS345) on 26 August 1949 off Norway. On 13 June 1950 the British submarine *Trenchant* suffered a shattering battery gas explosion. The submarine had been dived for nearly ten days, schnorkelling at night, while participating in the annual 'Summer War', and demands on the battery had been heavy with a high charge/discharge cycle. The explosion gutted the area between the control-room and torpedo stowage compartment and needless to say, *Trenchant*'s part in the exercise was over. Following this accident Flag Officer Submarines felt compelled to comment that, 'Like ammunition it [schnorkelling] is safe only so long as one remembers that it can be VERY dangerous.'[2]

Undoubtedly the most famous schnorkel casualty was the Israeli submarine *Dakar* which sailed from the UK for Israel in January 1968. The passage to Gibraltar was to be made on the surface, but that from Gibraltar to Haifa was to be made submerged, schnorkelling by night and diving deep by day. Two minutes after midnight on 25 January 1968 *Dakar* made her routine signal giving her position as 300m west of Haifa. That was the last anyone heard of the submarine. When her next two signals were not received an intensive air and sea search was launched but on 31 January the submarine was formally declared lost.

What had happened? *Dakar* sailed heavily overloaded with stores, with a comparatively inexperienced crew and against the advice of the Royal Navy who considered that the crew needed a proper work-up. The most likely explanation for her loss is that hydrogen gas built up to dangerous levels during that long submerged transit of the Mediterranean. The crew lacked the experience to appreciate what was happening, and it would have taken only one small spark from an electrical circuit or appliance to cause a massive and catastrophic explosion.

The schnorkel was really little more than a temporary solution to the problem of diesel-electric submarines having to surface to charge their batteries. Schnorkelling still exposed the boat to detection. What was required was a powerplant that would allow the submarine to remain underwater without the need for surfacing. It was the Germans who took the first steps in this direction. Dr Helmut Walther developed a closed cycle propulsion system that used the thermal energy produced by the decomposition of hydrogen peroxide (HTP) to produce steam and oxygen at high temperatures (1,765 °F) which were fed into a combustion chamber to ignite fuel oil. For submarines this system gave all the benefits of high underwater speed. Initial trials in *U792* and *U794* were successful, with *U794* achieving an underwater run at 20 knots for five-and-a-half hours! However, the nightmare bureaucracy of the Third Reich and scepticism among engineer officers meant that at the end of the war only seven of the Type XVII Walther submarines were completed, and none of the Type XVIII long-range boats or the Type XXVI coastal boats.

After the war Britain and America dabbled in Walther turbines. Britain commissioned a Type XVII U-boat as HMS *Meteorite* and then built two unarmed experimental HTP submarines, *Explorer* and *Excalibur*. The two submarines were known as *Exploder* and *Excruciator* on account of their volatile performance. British wardrooms were alternately thrilled or appalled at the stories of the exploits of these submarines. On one occasion a dangerous build-up of carbon monoxide in *Explorer* while at sea resulted in the entire ship's company standing on the casing with every hatch open while the boat was ventilated. HTP is an extremely volatile substance and although Walther propulsion systems offered many advantages, it was quickly superseded by nuclear power.

Before discussing nuclear power the subject of air independent propulsion (AIP) must be mentioned. Despite the failure of HTP to provide a safe non-nuclear powerplant, there has been considerable development in this field. Some designers hope to produce electricity direct from fuel cells while others rely on external combustion engines (ECE). The most promising developments in this field come from Sweden where in 1987 two V4-275R Stirling engines were fitted to *Nacken* which was completing at Kockums yard. The Stirling engine operates by applying external heat to a sealed chamber which encloses a 'working gas'. A piston is used to alter the volume occupied by this gas while a second piston, called the Displacer, moves it from the hot to the cold part of the engine. External combustion is continuous, as is the movement of both pistons. The engine provides almost constant torque and a very quiet and smooth performance ideally suited to the submarine environment. The only disadvantage to the system is that it creates a CO_2 exhaust. This can be

HMS Excalibur, *one of two experimental British submarines built after the war which used hydrogen peroxide in a closed-cycle engine. Though it promised much, HTP ultimately gave way to nuclear power. (MoD)*

discharged with the boat as deep as 300m but it does create a detectable heat signature and noise. Stirling engines are now fitted in Swedish Gotland-class SSKs while Australia has the option to retro-fit AIP in the RAN's Swedish-built Collins-class SSKs.

It was nuclear power that gave substance to the submariners' dream for unlimited underwater endurance. The nuclear submarine is unique in combining underwater and long-range surface propulsion because its fuel does not require oxygen for combustion. Nuclear submarines are also unique in that the powerplant produces so much power that there is sufficient for many auxiliary functions which would be beyond the capacity of a diesel boat. These include sophisticated active and passive sonars along with their computer support, health-physics plants that allow the boat to renew its atmosphere every 24 hours, and a host of other functions.

The United States had developed British theoretical research into splitting the atom with the intention of producing a war-winning weapon. After the war further applications were found for this almost limitless source of energy. One of these was to power ships and submarines. An American naval officer, Capt Hyman G. Rickover, an engineering specialist

USS Nautilus, *the world's first nuclear-powered submarine. Note the striking resemblance of the hull to the German Type XXI design. (US Navy)*

whose career to date had not been spectacular, became involved in the work and single-handedly delivered nuclear power to the US Navy. His admirers claimed that 'Rickover saved nuclear power for the Navy' while his detractors claimed that 'Nuclear power saved Rickover for the Navy'. Rickover was a complex character about whom opinions are seldom neutral. He was abrasive, opinionated and rude. He used every trick in the book in order to maintain his position as sole custodian of the atom for the US Navy and dealt ruthlessly with all interlopers. However, despite criticisms of the man and his regime, it cannot be denied that he did deliver nuclear power to the US Navy and with it a 100 per cent record for safety and engineering excellence.

USS *Nautilus* was the first nuclear-powered submarine. She employed a pressurised water reactor (PWR) of the type which has formed the basis for most Anglo-American development since then. In a PWR water coolant passes round a primary circuit and through the reactor pressure vessel where it is heated by the nuclear fuel elements before passing through the steam generator. High pressure is maintained in the primary circuit to prevent the water from boiling. In the steam generator heat from the primary circuit converts water to steam which is then used to drive a conventional steam turbine. From there a system of clutches, gearing and propulsion transmits the power to the sea. Steam is also used to drive the turbo-generators which supply the submarine with electricity. Although the PWR has remained the mainstay of submarine nuclear power, both America and Russia have experimented with liquid metal coolant which heats the water in the steam generator more efficiently. America's flirtation with liquid metal coolants ended in 1960 when the liquid sodium plant was removed from *Seawolf* on the grounds that it was unsafe. The Russians have persisted with liquid metal designs in the Alpha, Sierra and Mike classes.

The main problem with nuclear-powered submarines – a problem that was only appreciated when the first examples put to sea – was how noisy they were. Nuclear machinery is inherently noisy. Even at very low speeds the circulation pumps that kept the primary circuit running and the turbo-generators could not be turned off. Turbine reduction gearing added its own distinctive signature. The main solution to the problem has been in the development of PWR reactors where the noisy circulation pumps are replaced by natural circulation of both the primary and secondary circuits. The first such reactor first went to sea in 1969 in the USS *Narwhal* and proved very successful.

No discussion of propulsion is complete without some reference to noise reduction. If a submarine can be heard, it can be detected, tracked and destroyed. Noise reduction is as important a subject to the submariner as camouflage is to the soldier. It was not always thus. The subject of noise reduction has only developed parallel with the development of long-range passive sonars. In other words, as the means of detection have been developed, so submariners have sought to reduce the chance of their being detected.

During the Second World War the problem of noisy machinery was appreciated though not always dealt with. The French submarine *Narval* was so noisy that she 'had to be heard to be believed',[3] while a British submarine commander described his Freon air-conditioning pump as making a noise akin to 'the wrath of God'. Submariners knew that they had to switch off noisy equipment when attacking or being attacked, but little thought was given to noise reduction in the design or construction process. The specification for a 1930s British T-class submarine does not mention the subject at all.

Today the subject is rightly treated almost as a religion among all submariners. The simplest method of noise reduction is to keep noisy machinery switched off. When dived, submarines usually maintain one of a series of 'quiet states' which introduce progressive controls on what equipment can be operated. The quietest is ultra quiet state (UQS) in which all equipment is turned off except for that necessary to maintain control, support life and service passive sonars. UQS is extremely unpleasant for all concerned, particularly in the tropics. The commander has to decide whether elimination of noise by switching off a particular item of equipment is justified by the subsequent decreased efficiency of the crew.

Better noise reduction is achieved at the design and construction stage by fixing individual items of equipment on resilient mountings. An alternative approach is to mount whole groups of equipment, such as turbo-generators, on 'rafts' with flexible connections between machinery on and off the raft. Such a raft can be 'locked' when the boat is operating at speed. However, care has to be taken to ensure that careless construction and/or maintenance does not create 'noise shorts' between the raft and the hull. When HMS *Warspite* was examined in 1969 following a sound range test, it was found that unlocking the raft made no difference to the boat's noise signature! Subsequent and exhaustive examination revealed no fewer than twenty-one direct noise shorts between raft and hull – nearly all the work of one fitter who had not appreciated the nature of the pipework he was connecting. Acoustic housekeeping is now an extremely complex subject. We are all familiar with Second World War films where the submariner dare not even move for fear of making a noise. Developments in insulation mean that a modern submarine is so well insulated that the noise of a grenade explosion inside would not escape the hull.

What future developments can be expected in terms of submarine propulsion? AIP plants will doubtless gain in popularity among those nations still running SSKs. Nuclear plants will probably get smaller while great efforts will be made to reduce noise by cutting down on the number of pumps and associated machinery. Bigger and more effective electric motors may well replace turbines. Attention has also been paid to reducing water resistance around the hull by, in effect, making the hull 'slippery' so it passes through the water more easily. The US Navy has experimented with 'lubricant' dispensers in the ballast tanks of some 688 type SSNs. It is the Russians, typically, who have gone a step further and developed a system on the SSN *Beluga* whereby a polymer dispenser at the bow gives the boat a false 'skin' over which the water flows with much less resistance. Figures are unavailable but speed is certainly increased while at the same time the polymer acts as acoustic insulation, further reducing engine and reactor noise from within the submarine.

NOTES

1   Werner, Herbert, *Iron Coffins* (London, Arthur Barker, 1970), p. 299.
2   Submarine General Letter, 20 September 1950, RNSM A1949.
3   Report by Captain(s) One on *Narval*'s loss in December 1940. RNSM Archives.

LIFE IN A STEEL TUBE

I will end by giving you an expression of opinion about life in general in submarines. I always feel it is a strange thing to like, but if you do like it, which most do who take it up, you like it very much indeed.

Adm Sir Reginald Bacon

Aft of the accommodation spaces is the control-room, the nerve centre of the boat. From this compartment ladders lead up to the bridge. The control-room contains the helm, the hydroplane operators' positions, periscopes, chart tables, blowing panel and all the other important command instruments of the submarine. Around the control-room are grouped important offices such as those for wireless, Asdic and radar. It is in the control-room that differences in national practice become very apparent. In British submarines the commanding officer directs both normal manoeuvres and attacks on enemy shipping from this compartment. While he does so the first lieutenant keeps an eye on the planesmen and is responsible for the trim, the level of horizontal stability, of the submarine.

British COs were thus under the interested, and informed, scrutiny of their ships' companies at all times. The lower deck was quick to spot a non-performer and morale suffered accordingly. However, in German U-boats and French and American submarines (and some, though not all Japanese submarines) a second compartment was situated directly above the control-room. Sometimes known as the conning tower, this compartment held the monocular attack periscope and attack instruments. When conducting a dived attack, the commanding officer would climb up to this compartment with his key ratings to conduct the attack, leaving the first lieutenant or OOW down in the control-room to manoeuvre the boat according to his orders. One American CO, Lt Cdr 'Mush' Morton of *Wahoo*, let his XO, Lt Richard O'Kane, conduct the attacks while he conned the boat from the control-room – a striking demonstration of trust in a subordinate. When O'Kane got his own boat this was one practice he did not continue. Inside the control-room another key difference showed itself. In British, American, French and Japanese boats engineeer officers were confined strictly to the engine room. Certainly in the Royal and Imperial Japanese Navies, engineer officers lacked any executive authority. The US Navy had no such social divisions but even so engineers officers were expected to stay where they belonged – back aft. However, in the

Opposite: *An Austrian officer sits in the control-room of his submarine writing up the day's events. However hectic a day might have been, there was always paperwork to attend to. (Author)*

The hydroplane operators' position in a British wartime submarine. Such positions were common to all submarines. Note the depth-gauge set at 5ft, indicating that the scene was set up for the benefit of the photographer!

Reichsmarine and *Kriegsmarine* the place of the engineer officer, the *Leitender Ingenieur*, was in the control-room. Here he supervised the planesmen and was responsible for the trim – duties which in other navies would be carried out by the first lieutenant or officer appointed for the purpose. In a German U-boat the *LI* was very much the CO's right-hand man. The first lieutenant, the 1WO, played hardly any part in the dived operations apart from normal watchkeeping. His responsibility was the working of the bridge-sight during torpedo attacks carried out on the surface.

Above the control-room is the bridge which is built up from the casing. The bridge contains basic navigational instruments for surface navigation and mountings for close-range weapons. The 'tunnel' which runs down from the bridge to the control-room is fitted with hatches at its upper and lower ends. Even if those on the bridge are killed and/or wounded and are unable to secure the hatch, then the lower hatch can be secured and the submarine dived. The tunnel was also the main means of drawing air into the boat before the advent of the schnorkel. The passage of an unusually bulky member of the ship's company could cause a temporary vacuum in the boat. (For instance the commanding officer of the British submarine

German seamen carry out the daily maintenance routines on torpedoes in the fore ends of their submarine during the First World War. This task still continues unchanged despite the huge developments in torpedo armament since then. (U-boot Archiv)

Turbulent, Cdr J.W. Linton RN, was of somewhat rotund build and the ship's company always knew when he was going to or from the bridge because of the two vacuums created as he passed through the upper and lower hatches.) In an emergency dive everyone had to get down through the hatch and down to the control-room as quickly as possible, regardless of crushed fingers, toes or egos.

Some British submarines, such as Porpoise-class minelayers and early S- and T-classes had a 'cab'-type bridge fitted which gave the watchkeepers some protection from the elements. These cabs were a relic of peacetime when a boat might expect to spend some time on the surface. However, in wartime all-round visibility was more important and the cabs went.

Aft of the control-room there were more accommodation spaces along with the galley and heads for junior and senior rates and officers. Then came another watertight bulkhead which separated the engine-room from the rest of the boat. British, French, German and Japanese submarines had just one engine-room in which the diesels were arranged side by side. American fleet submarines had two engine-rooms, one aft of the other. Walking through a wartime US fleet boat gives the visitor an uncanny sense of *déjà vu*: having walked through the engine-room you come to another compartment which is exactly the same. Inside the engine-room is a hatch, larger than any of the other hatches on board, since it allows bulky engine-room parts/spares to be brought into/removed from the boat.

The bows of the British T-class submarine Totem *showing the arrangement of the torpedo tubes and how the requirement for a six-tube bow salvo dictates an oval shape to the pressure hull. (Capt John Coote RN)*

The control-room of a British E-class submarine looking aft. The hydroplane operators' positions are seen on the right-hand side of the photograph. (Author)

An officer in the Free French submarine Rubis *takes a bearing from the bridge. (ECPA)*

*The fore ends of the British
HMS* Rorqual *in 1941 looking aft.
Such spaces were typical of
submarines of all nations. (Author)*

Since this hatch represented the biggest single weakness in the pressure hull, it was usually secured while at sea with special strongbacks. In some British submarines this hatch also doubled as an escape hatch. Aft of the engine-room lay the motor-room, or manoeuvring-room in USN parlance. At the very after end of the boat was either the after torpedo room (in American and German submarines) or more accommodation space. Having the stokers and engine-room personnel accommodated aft was a sound decision. It meant that at diving stations or when the watch changed the first lieutenant/diving officer/LI did not have to worry about a large number of men moving from one end of the boat to the other and upsetting the trim. By common custom submarine COs usually forbear to examine too closely what is happening at the after end of their submarine. One British CO recalled that, 'The stokers liked being on their own back there – what they got up to God alone knows.'[1]

NOTE

1   Vice Adm Sir Hugh Mackenzie to author, 7 February 1989.

A game of 'Uckers', a variant of Ludo, played on the after casing of a British submarine in the Far East during the Second World War. Clearly the risk of air attack is very low! (Author)

THE TORPEDO

The only use of Holland *is to discharge torpedoes and no weapon is more erratic!*

Adm Charles O'Neill USN

Numerous submariners, past and present, would agree with O'Neill's assessment of the purpose of the United States Navy's first submarine in 1900. The torpedo has been, and remains, the submarine's principal means of attacking other ships and submarines. The torpedo began its life as a powered version of the naval mine and was first used during the American Civil War. Naval engineers attached explosive charges or torpedoes to long poles or spars, which could be thrust at enemy craft from small boats or submersibles. The first propeller-driven torpedo was perfected in 1864 by the English engineer Robert Whitehead. The Whitehead torpedo, with various modifications, was subsequently adopted by all the navies of the world.

Torpedoes were discharged from tubes usually situated at the bow and stern of the submarine. Some British E-, G-, J-, K- and L-class submarines also had torpedo tubes mounted athwartships facing port and starboard to provide a beam salvo. As submarines became larger, the number of torpedo tubes which could be mounted increased. In 1901 the British *Holland 1* carried one 18in torpedo tube; seventeen years later the R-class carried six 18in bow tubes. The desirability of fitting a large number of torpedo tubes was the subject of much debate. On the one hand was the argument for fitting the maximum number of tubes in order to have a large salvo of torpedoes for firing; on the other hand a large number of torpedo tubes could not be contained within the circular cross-section of the pressure hull. Thus in the Second World War the Germans settled for an armament of four 21in torpedo tubes in their immensely successful Type VIIC U-boats. These boats had a circular cross-section to their hulls for nearly their entire length and thus were very strong and could absorb a good deal of punishment. In contrast, most British and US Second World War submarines had an armament of six or more torpedo tubes which required an oval-shaped bow, thus weakening the hull structure.

One way of increasing the torpedo armament without fitting extra torpedo tubes was to carry the torpedoes externally – outside the pressure hull – either in launching cradles or in external torpedo tubes. French and Russian submarines of the First World War made extensive use of the Drzewiecki collar, or launching cradle, in which a torpedo was carried in a frame secured to the outside of the pressure hull. To launch the torpedo, the cradle was released from inside the pressure hull and the weapon

angled out to an approximate firing course. A French Pluvoise-class submarine was fitted with only one torpedo tube but could carry another six weapons in Drzewiecki collars. However, the theoretical advantage this gave was almost completely lost by the difficulties associated with the system. Torpedoes fired in this fashion could not be given a steady course, they were extremely vulnerable to weather damage and they limited the boat's diving depth to 30m.

An alternative was to fit torpedo tubes into the submarine's casing but outside the pressure hull. Nearly all French submarines of the Second World War carried an external twin or triple torpedo tube mounting, very similar to the type used in destroyers, fitted into the upper deck casing. This could be trained to fire on the port or starboard beam. British T-class submarines had four, and later five, single torpedo tubes built into their casings, increasing the armament by nearly 50 per cent without having any effect on hull structure or strength. In both the British and French external tubes, there was no possibility of maintaining the torpedoes at sea, nor could the tube be reloaded, so these weapons were usually fired first.

A torpedo tube is sealed at one end by a bow cap and at the other by the rear door. It was obviously vital that both should not be open at the same time. Indicators were fitted to show the position of the bow cap which was electrically or hydraulically operated in all but the earliest boats. Later interlocks were fitted which made it impossible to have both doors open at the same time. Loading and reloading operations were, and still are, supervised with great care: a 21in torpedo tube will admit water to a submarine at a rate of 2.25 tons per minute. Memories of the

HMS *Thetis* disaster in June 1939 (see page 123) are still fresh and painful after even after sixty years.

Until 1945 torpedoes were ejected from the tube by compressed air. Systems used during the First World War allowed the air to escape out of the tube along with the torpedo, thus giving away the firing submarine's position. The next stage of development was to 'suck back' the air as it reached the muzzle of the tube. The Germans developed an alternative method, using a piston that pushed the torpedo out the tube without the tell-tale air bubble. Even though it proved possible to prevent the air leaving the tube, the use of air as the means of ejecting the torpedo limited the depth at which weapons could be fired. It was also extremely noisy. This was not so important in the years leading up to 1945 when, generally speaking, torpedoes were either fired while the submarine was at periscope depth or on the surface. However, when ASW became an important submarine function in the postwar period it became clear that alternative means of torpedo propulsion were required.

One method was 'swim out', in which the torpedo propelled itself out of the tube. Space had to be provided around the torpedo to allow water to flow back into the tube as the torpedo moved forward. Thus the US Mk.37 ASW torpedo used during the Second World War was only 19in in diameter to allow it to swim out of a 21in diameter tube. The first practical swim out installation was developed for the German Type XXIII U-boat. Swim out requires that the torpedo is dynamically stable, as it leaves the tube at a very low speed, 8 to 10 knots, before accelerating away. Moreover, the bow of the submarine must be designed to keep hydrodynamic interference to a minimum so that the torpedo is not deflected as it leaves the tube – hence the wide bulbous bow shape adopted by German IKL-built submarines in the postwar period. Swim out tubes are compact and simple, can be used at any depth and are extremely quiet. However, they cannot be used to launch any non-propelled munitions such as mines or guided missiles which must be ejected from the tube before their propulsion systems start to run. Postwar German *U-boote* engaged in minelaying work carried their mines in external canisters bolted on to the casing. A more important disadvantage is that some kind of positive ejection is essential if a submarine has to fire while moving at speed.

A third method of ejection is water pulse ejection which is currently used by Britain, the United States and Russia. Before firing, the tube is flooded, thus equalising water pressure around the torpedo. Then a water pulse is 'fired' into the rear of the tube by a hydraulic pump or an air turbine pump, thus forcing the torpedo out of the tube. The system can be used at any depth but is very complicated to fit. To fire multiple salvoes each torpedo tube must have its own firing pump. French submarines are unique in using hydraulic rams to eject torpedoes.

Firing any torpedo (or mine or missile) causes an imbalance in the submarine's trim which is only partially offset by allowing the empty torpedo tube to fill with sea-water. It is essential to keep control of the submarine's trim after firing to prevent the boat from broaching and thus giving away her position. On 3 October 1918 the British submarine *L.10* successfully attacked and sank the German destroyer *S.33* in the North

Sea. However, after firing, her conning tower broke surface in full view of *S.33*'s consorts who sank her with gunfire. Consequently, all discharge systems have to include means for compensating for any net loss of weight. The usual means of achieving this, known as Automatic Inboard Venting (AIV) is to allow the tube to remain open long enough after firing to allow an extra 350 litres of water to flood into a special AIV tank.

There are five parts to a modern torpedo: warhead, powerplant, sensors, control surfaces and umbilical to the firing submarine. Until the middle of the Second World War torpedoes consisted of only the first three, with a warhead consisting of high explosive actuated by a contact fuse, a simple gyro to keep the torpedo on a steady course and a powerplant. The warhead was composed of either TNT or a much more powerful explosive called Torpex. Detonation was achieved by a simple contact fuse or later by a magnetic fuse which detonated the warhead as the torpedo entered the target's magnetic field. An explosion beneath the target did far more damage than a simple hole blown in the side. Damage control could effectively localise the effects of a hole blown in the ship's side, but an explosion under the ship would do massive structural damage which might result either in total structural collapse or in the damaged target having to be towed back to base by other vessels (whose presence might be required elsewhere).

The torpedo was kept on course and at the correct depth by a simple gyro. Early torpedoes could not be angled to a pre-set firing course so the submarine itself had to be pointed at the target before firing. American and German commanding officers were given more freedom of action since their submarines were fitted with fire control instruments that allowed a specific firing course to be applied to the torpedo's gyro while it lay in the tube before firing. These instruments also calculated a continuous firing solution which meant that the machine did not have to be reset at every periscope observation.

During the Second World War the Germans and the Americans suffered significant problems with torpedo reliability. A common cause for this was the lack of experience in the live firing of torpedoes during peacetime. It was, inevitably, expensive to carry out such exercises, but there were other factors which weighed in against exhaustive trials: scientific and technical claims about reliability, commercial interests, the exclusion of submariners from the decision-making process and secrecy.

Torpedo faults come into two main categories. First, the torpedo runs too deep and thus roars harmlessly underneath the target it was meant to strike. Secondly, the fuse or exploder in the warhead does not detonate the weapon either on contact or after entering the appropriate influence field. In addition, of course there is the occasional torpedo that runs 'rogue' after being fired and presents a hazard to friend and foe alike! However, given the conditions in which torpedoes were stored and maintained on board a submarine, it is a wonder that any of them worked at all. A torpedo was, and still is, an immensely complex piece of machinery which requires daily maintenance routines.

These routines were carried out in an impossibly crowded space, where a number of men were trying to sleep, eat and generally relax, with the boat often pitching and rolling according to the sea state. This work was

The power of the torpedo. The Italian cruiser Muzio Attendolo *following an attack by the British submarine* Unbroken *in August 1942 in which she lost her bows. (Dott Achille Rastelli)*

usually done by senior rates, whose contribution to the success or otherwise of their submarine cannot be overestimated. A secondary factor was that the submariners had to take what was issued to them by the depot ship/submarine shore base. Sloppy work by either of these organisations could result in poor performance at sea. It was noticeable that torpedo reliability in the United States Navy went up dramatically during the Second World War when the command insisted that everyone who had a hand in torpedo maintenance at any stage, from factory to submarine, should sign for any work they had done. Personal accountability proved a remarkable incentive.

Even so, mistakes in torpedo preparation and maintenance could happen in the best regulated navies. Stop valves were not opened, safety devices were left in place and, on occasions, propeller guards were left *in situ*. However, a series of torpedo failures in the early stages of the war convinced Adm Dönitz that something other than poor attacking techniques or sloppy maintenance by an *Oberbootsmann* was wrong with his torpedoes. In the case of the *Kriegsmarine* torpedo development was in the hands of the *Torpedoversuchsanstalt* (TVA) which designed and developed the weapon and then issued it to the submariners. In other words, there was little contact between the designers and the users.

On 19 September 1939 Kl J. Franz in *U27* was mortified to watch his torpedoes explode prematurely in front of a Tribal-class destroyer. On 30 November 1939 ObLtz.S Gunther Zahn of *U56* penetrated a screen of twelve British destroyers to attack the battleship *Nelson*, only to hear the torpedoes strike but not explode.[1] KL Gunther Prien of *U47* complained to Dönitz that he could 'hardly be expected to fight with a dummy rifle!'[2] Prien spoke for all submariners: they were fighting the war at sea and did not want to see their attacks fail because of errors by shore-based designers and maintainers. Finally, Dönitz complained to Raeder, the commander-in-chief.

An inquiry was commissioned which concluded that there had been negligence by officials of the TVA, four of whom were subsequently court-martialled. The inquiry also discovered another reason why German torpedoes were running too deep. When a submarine is dived the inevitable leaks from HP air lines cause the air pressure in the boat to rise. The balance chamber in the torpedo was not designed to resist such a rise in pressure over a period of time, which might be as long as a few weeks, if the boat was on a long patrol. When the torpedo was fired, the depth-keeping hydrostatic valve was already biased, taking the torpedo deeper than had been set. No better example can be found of the results of failing to test a weapon under operational conditions.

However, the problems suffered by the *Kriegsmarine* were as nothing compared to those suffered by the US Navy. This was nothing less than a crisis of confidence in their equipment by US submarine commanding officers. From the outbreak of war reports came in of torpedoes missing the target even though they were running 'hot, straight and normal'. The torpedo most widely used by the USN was the Mk.14, although older S-class submarines used the smaller Mk.10. The Mk.14 weighed 3,280lb and had a range of 4,500yd at a speed of 40 knots, or 9,000yd at 36 knots, with a warhead of 635lb of Torpex. The torpedo was fitted with the Mk.6 exploder, an extremely complicated mechanism that was designed with a dual contact and magnetic setting. The latter was considered particularly important for attacks against warships where an explosion under the keel would have a much greater effect than multiple hits on the warship's stoutly protected sides. The Mk.6 had been designed and tested under conditions of such secrecy that only submarine commanders and their torpedo officers were authorised to know of its existence. Almost as soon as the first US submarine put to sea on a war patrol in December 1941 there were reports that torpedoes were running too deep or that the Mk.6 exploder was not performing. However, the command refused to listen to the COs' complaints: commanders who persisted were often given an official reprimand for daring to question official wisdom. However, in April 1942 Rear Adm Charles Lockwood took over command of the Asiatic Fleet submarines and at last began to take note of what his COs were saying. A memorandum from Lt Cdr John Coe of *Skipjack* made the problem crystal clear: 'To make an 8,500 mile round trip into enemy waters, to gain attack position undetected within 800 yards of enemy ships only to find that the torpedoes run and over half the time fail to explode, seems to me an undesirable manner of gaining information which might be determined any morning within a few miles of any torpedo station in the presence of comparatively few hazards.'[3]

Lockwood carried out his own tests using *Skipjack* and a sister submarine *Saury* and reached the astonishing conclusion that the torpedoes ran 11ft deeper than set! He now had to do battle with the Bureau of Ordnance (BuOrd) who were not going to give in without a fight. A number of officers had staked their reputations on the Mk.14/Mk.6 combination and to some this was more important than simply fighting the war. After their initial refusal to accept Lockwood's data, the Bureau of Ordnance finally caved in after the redoubtable Adm 'Ernie' King, the Chief of Naval Operations, demanded an explanation. The Bureau then shamefacedly organised their own trials using USS *Herring,* which found that the Mk.14 was running some 10ft deeper than set owing to *improper testing* [author's italics].

The problem of torpedoes failing to hit the target seemed to get worse. USS *Sargo* reported thirteen consecutive hits but no explosions, while *Seadragon* had fifteen such duds. The finger of suspicion was pointed firmly at the Mk.6 exploder. Lockwood had now moved to Pearl Harbor as ComSubPac, his place in Fremantle being taken by Capt Ralph Christie, a torpedo specialist who had been closely involved with the 'testing of the Mk.6 exploder'.

On 11 June 1943 USS *Trigger* (Cdr Roy Benson USN) fired four torpedoes at the Japanese carrier *Hiyo* in Tokyo Bay. Benson claimed *Hiyo* as 'sunk' after hearing four explosions. However, on his return to Pearl Harbor, Lockwood told him that only one of his torpedoes had exploded; the other three had hit without detonating. *Trigger*'s failure to sink *Hiyo* was the last straw for Lockwood. On 24 June 1943 he ordered all submarines under his command to de-activate the magnetic settings on the Mk.6. But down in Fremantle, Christie took no such action. This divergence in opinion placed US submarine commanders in an awkward position. When they worked for Lockwood they could de-activate the magnetic setting on the Mk.6. but as soon as they crossed the boundary into Christie's area, they had to re-activate them. Christie was ruthless in his dismissal of any criticism of 'his' Mk.6. When Lt Cdr 'Moke' Millican of USS *Thresher* complained that after an attack on a Japanese submarine his torpedoes merely 'clinked 'em with a clunk', he was relieved of his command and returned to the United States for a rest and appointment to a new construction. The ultimate story of the failure of the Mk.6 concerns USS *Tinosa*'s (Lt Cdr Dan Daspit USN) attack on the factory ship *Tinosa Maru* on 24 July 1943. Daspit fired no fewer than fifteen torpedoes at *Tinosa Maru* and recorded fifteen hits. He could even see the small splash thrown up as the weapon struck the ship, but not one of the torpedoes actually exploded. Disgusted, Daspit kept his sixteenth and last torpedo for examination back at Pearl Harbor. What the crew of *Tinosa Maru* made of it all has not been recorded.

When Daspit returned to Pearl Harbor and reported to Lockwood, the latter decided to dispense with any further correspondence with the Bureau of Ordnance. Lockwood ordered a firing trial from *Muskallunge* against the submerged cliffs at Kahoolawe. When one of the torpedoes failed to go off, it was recovered, gingerly, for examination. It was found from this experiment, and other trials, one of which involved dropping a torpedo nose first (but without a warhead) from a dockyard crane, that a minute modification had to be made to the size of the firing pin – a modification that could be managed in any machine shop at Pearl Harbor. On 30 September 1943 USS *Barb* left Pearl Harbor with an outfit of twenty torpedoes, each fitted with the new firing pin. It had taken nearly two years for US submariners to be given a reliable torpedo.

Even the Royal Navy was not immune from these problems. In 1942 the British introduced a magnetic firing pistol, the compensated coil rod (CCR), to supplement the traditional contact firing pistol. Almost as soon as it was introduced reports came in of torpedoes failing to explode. Perhaps the most mortifying failure of CCR occurred when HMS *Thule* (Lt A.C. Mars DSO RN) fired three torpedoes against a Japanese R0-100-class submarine off Penang. When Mars raised his periscope he was rewarded by the sight of a column of water in the air and the submarine's stern rising nearly to the vertical. *Thule*'s ship's company jubilantly sewed a 'U' on their 'Jolly Roger', but postwar analysis revealed that what Mars had actually seen was the premature explosion of one of his CCR-fitted pistols and the Japanese boat performing one of the fastest crash dives in submarine history. In 1944 the Admiralty withdrew all CCR pistols.

Early torpedoes were powered by compressed air and their performance depended on pressure built up in air flasks. Later, heaters were added to boost air pressure, and by the Second World War the standard method of propulsion was steam, or rather steam-boosted air. It was a fairly straightforward system, although the torpedoes required thorough daily maintenance while at sea, but it had the disadvantage of leaving a clear wake of bubbles on the surface. Not only did this wake give the target a chance to manoeuvre away from the torpedo's track but it meant that a determined counter-attack could be mounted by running down the track of bubbles to the firing point.

During the 1930s several navies sought major improvements in torpedo performance: greater range, greater speed and greater warhead size. Improvements in all these areas were frustrated by the fact that the torpedo's size was governed by the fact that it had to fit into a torpedo tube whose size was fixed. Examples of development include the Japanese introduction of oxygen-driven torpedoes – essentially wakeless – and American experiments with HTP. Several navies experimented with electrically driven torpedoes but it was the Germans who first employed an electrically driven torpedo at sea. Later in the war the US Navy adopted electrical propulsion for the Mk.18 torpedo in the Pacific. However, the advent of electrical propulsion was not without its problems. The first torpedoes had endless problems: 'The first ones taken to sea lacked hydrogen burning circuits and therefore had to be frequently withdrawn from the tubes for ventilation. Several instances of hydrogen explosions and fires occurred. One fire on board *Flying Fish* heated the warhead until the torpex melted and ran out.'[4]

Perhaps the most significant development during the Second World War was the introduction of the homing torpedo. This was initially a German invention and reflected their desire to regain the initiative in the Battle of the Atlantic against stiff Allied counter-measures. The T5 *Zaunkonig* torpedo (known to the British as GNAT) was fitted with directional transducers in the nose that turned the rudders to steer the torpedo to a noise source, usually the target's propellers. It was designed primarily for retaliation against escorts who were now forced to tow astern acoustic decoys (Foxers or CATs), which seriously impeded their own performance. A further German development was the pattern-running torpedo, FAT (*Feder Apparat Torpedo*), which ran in a series of loops after firing, and

Torpedo fire control in a British submarine of the Second World War period.

LUT (*Laguenabhangigertorpedo*) which could be fired independently of the target's inclination. These two weapons were supposed to be able to allow U-boat commanders to fire at a convoy from some distance without the attendant risks of discovery, counter-attack and destruction. They were less successful than Dönitz hoped and in any case, the weapons were available only in tiny numbers at the time when they were most needed.

In the postwar period torpedo development advanced several stages, driven wholly by the emphasis on submerged ASW. Torpedoes now had to be able to chase a target that was capable of manoeuvring radically at great speed in the vertical and horizontal planes. This meant that the torpedo had, literally, to think for itself or rely on information transmitted from the firing submarine via an umbilical. The torpedo had to be able to change depth in response to the commands of the firing submarine and run at speeds greater than the predicted top speed of its opponent.

This scale of development has not been easy. Rapidly changing technology, the changing perceptions of the threat and budgetary constraints have all combined to make torpedo design and procurement a nightmare process. The American Mk.48, which is the mainstay of the US submarine armament in 1998, actually began life in 1957 as the EX-10! It took seven years before the requirements were finalised in 1964 and it was not until 1972 that the first class trials began. In the intervening years there have been a series of intermediate weapons and upgraded older ones.

British efforts in this area are even more lamentable. The Mk.XXIII wire-guided torpedo was tested in 1955, ordered in 1959 but not fully cleared for front-line use until 1971! In the meanwhile the Royal Navy, which was spending *billions* of pounds of taxpayers' money on nuclear-powered fleet submarines, relied on the Mk.VIII torpedo which first went to sea in 1937. Even when Tigerfish was cleared for front-line service, it was always considered unreliable – so much so that during the Falklands War the commanding officer of HMS *Conqueror* preferred to use the 'unintelligent' Mk.VIIIs to sink the *General Belgrano* rather than risk ruining the attack with the unreliable Tigerfish. The Russians have not admitted to any such problems but it is highly likely that their torpedo development path was equally painful.

The most important part of the modern torpedo is the 'brain'. This is situated in the nose and consists of a ring of microprocessors – the 'brain' of a British Spearfish torpedo has some 30,000 components. Its first task is to receive commands from the firing submarine until the point is reached when the torpedo is *enabled*, the warhead armed and homing circuits activated. The brain may also receive instructions to alter speed and depth and even to stop the torpedo in the event of the target being lost. The brain can then send back data to the submarine so that a fresh fire control solution can be produced. The second task of the brain is to operate the torpedo's organic active and passive sonar and to switch between these two modes as required. The brain can be programmed to ignore any decoy deployed by the target. In the unlikely event of a 'miss', the brain will organise a second attack, run and re-energise the firing circuits. The third function of the brain is actually to control the torpedo in its up/down and port/starboard movements by giving signals to the control surfaces at the stern of the weapon.

The size of the brain and the limitations on torpedo size have meant that torpedo warheads have actually decreased in size from the days of the Second World War. There is simply not enough room in a modern torpedo to pack in the brain, the powerplant and the explosive charge. In the 1950s the Americans sought to compensate for this by deploying a nuclear torpedo, the Astor. However, the deployment of nuclear torpedoes involved significant command and control issues and the weapon was dropped in favour of the conventional Mk.48. In 1945 the British Mk.VIII torpedo (and the US Mk.14) had warheads of 340kg and 300kg of Torpex respectively. The modern British Spearfish has a warhead of only 45kg and yet is equally lethal. This is because the Spearfish carries a directed-energy charge that will penetrate the target's hull like a hot steel bolt. Sea pressure will do the rest, tearing the hull apart.

Early ASW torpedoes were electrically driven and powered by high capacity silver zinc batteries. These were adequate to deal with most diesel-electric submarines. However, they were totally unsuited to deal with an SSN capable of 40 knots plus underwater. This requirement can only be met by employing non-stop closed cycle engines whose thermal energy is generated by a chemical reaction between a liquid metal fuel and a gaseous oxidant supplied from separate tanks. After the reaction begins, high pressure water is pumped into a mini boiler that produces steam which drives a turbine. Spent steam goes through a condenser and is returned to the boiler. Attached to the shaft is a generator that supplies the power for the 'brain'. A pump-jet propulsor is used, both to eliminate noise and to gain maximum efficiency. Control surfaces mounted in the efflux of the pump-jet make modern torpedoes like Spearfish, the US Mk.48 ADCAP and the Mk.50 Barracuda extremely agile.

Lastly, in order to exploit the considerable range of modern sonar, some kind of guidance from firing submarine to torpedo is desirable. This is best achieved by paying out a wire from the torpedo as it races away from the submarine. The torpedo can then be steered into the target until it is close enough for its own sensors to take over.

Table 1: Modern Wire-guided Dual Role Heavyweight Torpedoes

Nation	France	Russia	UK	USA
Model	F-17	53	Spearfish	Mk.48 ADCAP
Range (km)	20/25	20	40 or 65	38 or 50
Speed (kts)	40 or 35	46	70 or 60	55 or 40
Engine & Power	Electric	Steam	Turbine engine thermal	Piston Engine gas
Warhead (kg)	250	400*	45**	267
Weight (kg)	1360	1610	1850	c. 1600

* A 15KT nuclear warhead can also be fitted.
** Estimated figure: the 45kg warhead is a Directed-energy warhead.

The torpedo is still an important submarine weapon although it is now *primus inter pares* rather than enjoying a position of exclusive domination. Torpedoes now compete with self-propelled mines, tube-launched missiles and vertically launched missiles. Furthermore, the torpedo will continue to evolve. The Russians have deployed Shkval – a 200 knot jet-propelled torpedo while the Americans are reported to be working on a Tomahawk-launched torpedo. High-speed, short-range torpedoes are under development for anti-torpedo defence. The 'erratic' torpedo has come a long way since the early 1900s. Modern torpedoes can be as erratic as their predecessors but there is no sign of this weapon losing its position as the submarine's principal weapon.

NOTES

1   Zahn was incorrect in his analysis of the attack but his complaint was accepted.
2   Prien, G., *Mein Weg nach Scapa Flow*.
3   Blair, *Combat Patrol*, p. 56.
4   Roscoe, T., *United States Submarine Operations in World War II*, p. 262.

START THE ATTACK!

*F**k Me! I've hit it!*
Alleged comment by a British submarine commanding officer in
1940 on seeing his first wartime salvo strike the target.

O n 9 December 1912 Lt Giorgios Paparrigopoulos of the Royal
Hellenic Navy, commanding officer of the submarine *Delfin*,
became the first modern submarine commander to carry out a
torpedo attack. On patrol off the entrance to the Dardanelles
he sighted the Turkish battleship *Mejidieh* emerging with five escorts.
Paparrigopoulos managed to get within 500m of *Mejidieh* before firing a

The Italian cruiser Giuseppe
Garibaldi *sinks on 18 July 1915
after being torpedoed by the
Austrian* U4. *(Dott Achille Rastelli)*

The Russian submarine Morzh *makes a rare successful torpedo attack on a Turkish merchant ship in the Black Sea in 1916. (Museum of the Black Sea Fleet, Sevastopol)*

single torpedo. Unfortunately, the torpedo failed to run properly and sank. Over the next eighty-six years numerous submarine commanders of all nationalities suffered the same frustration at seeing a carefully set-up attack ruined by a faulty torpedo. Despite the failure of the attack, Paparrigopoulos's place in history is assured.

It fell to the German Navy to achieve the first sinking by a submarine. On 5 September 1915 the old cruiser HMS *Pathfinder* (Capt M. Leake RN) was on patrol off the Firth of Forth when she was attacked by *U21* (KL Otto Hersing). Standing Orders emphasised a minimum speed of 15 knots in waters where U-boats might be expected to operate, but *Pathfinder* was doing only 6 knots owing to the requirements of her task and her small coal bunkers. Consequently she was almost a sitting duck for Hersing. The torpedo was seen approaching the starboard side and helm was put over but it was too late. The torpedo hit the starboard side under the bridge and set off the forward magazine. The explosion blew Capt Leake aft and into the upper deck meat-safe. The bows were literally blown off and then *Pathfinder* started to sink as water poured into her hull. Four minutes later she had disappeared. There were only twelve survivors from her ship's company of 268.

Just over two weeks later there was another, more potent, demonstration of the submarine's ability. The cruisers *Aboukir* (Capt Drummond RN), *Cressy* (Capt R.W. Johnson RN) and *Hogue* (Capt W. Nicholson RN) were on patrol off the Dutch coast in order to prevent German forces from moving south to attack Channel troop convoys. Owing to bad weather on 20/21 September the three cruisers were without their destroyer screen. At 06.30 hrs on 22 September there was a violent explosion on *Aboukir*'s starboard side. She quickly assumed a 20° list to starboard while efforts were made to right the ship by counter-flooding. However, the list kept on increasing until it was clear that she was going to capsize. 'Abandon Ship' was ordered and *Hogue* prepared to rescue her crew. At 06.55 hrs *Aboukir* capsized and floated bottom up for a while before sinking. Capt Drummond thought that his ship had been mined but in fact she had been torpedoed by *U9* (KL Otto Weddigen).

As *Aboukir* was sinking Capt Nicholson took *Hogue* up to rescue her crew. *Hogue*'s crew threw anything that would float over the side to their comrades in the water. Almost immediately *Hogue* was struck by two torpedoes fired by *U9* on her port side and began to sink by the stern. At the same time the submarine broached on the cruiser's port quarter and she was immediately fired on by every gun that would bear. Ten minutes after being hit *Hogue* rolled over on to her beam ends and sank. It was now abundantly clear to Capt Johnson what had been responsible for sinking his two consorts. However, *Cressy* remained stationary picking up *Hogue*'s and *Aboukir*'s survivors. She thus became Weddigen's third victim in little less than an hour. Struck by two torpedoes, she sank equally quickly. A total of 60 officers and 777 men were saved from the three ships, which means that the death toll was 1,459 – greater than the total British casualties at the Battle of Trafalgar. Among the dead were many boy seamen from HMS *Ganges* and officer cadets from the Britannia Royal Naval College at Dartmouth.

The attack is significant for two other reasons: first, the sheer scale of the losses which one submarine was able to inflict on the British; and secondly, the total lack of any counter-measures. Unless Weddigen brought *U9* to the surface and exposed her pressure hull to gunfire or ramming there was nothing that the three cruisers could do by way of retaliation.

HMS *Barham* had left Alexandria on 24 November in company with other units of the battle fleet to cover a cruiser and destroyer operation against an Italian convoy. At 16.30 hrs on the following day she was attacked by *U331* (KL Franz-Dietrich Freiherr von Tiesenhausen). *U331* penetrated the destroyer screen with ease and fired three torpedoes at *Barham*. The submarine briefly lost trim and broke surface on firing but escaped, passing so close down the side of HMS *Valiant* that her guns could not depress far enough to engage her.

The impact of the torpedoes was clearly felt by Adm Cunningham in his flagship *Queen Elizabeth*. He rushed up to the bridge and saw, 'The poor ship rolled nearly over on to her beam ends, and we saw men massing on her upturned side. A minute or two later there came the dull rumble of a terrific explosion as one of her main magazines blew up. The ship became completely hidden in great clouds of yellowish-black smoke. . . . When it cleared away the *Barham* had disappeared.' The destruction of the ship was primarily attributed to a failure on the part of the screening destroyers: *U331* passed between two of these destroyers at comparatively high speed but neither destroyer made contact. (One destroyer's Asdic operator did make contact but was told to disregard the echo.)

Submarine vs submarine encounters are quite common. However, one of the parties involved has invariably been on the surface at the time of the attack. Submarine vs submarine encounters when both parties are dived are much rarer. Such attacks have taken place but usually more in hope than anything else and have been frustrated by the lack of proper sonar equipment or homing torpedoes. On 9 February 1945 *U864* (KL Rolf-Reimer Wolfram) was sunk by HMS *Venturer* in an attack in which both submarines were dived. *Venturer* (Lt J.S. Launders RN) was on patrol off Fejeosen (now known as Fedje), the most frequently used entrance to the important port of Bergen when, shortly after 09.32 hrs, she detected faint noises on her Type 129 hydrophone. The noise steadily increased, and at 10.50 hrs its source was confirmed as being a submarine when the OOW sighted a periscope on the HE bearing. The submarine was not schnorkelling as the tell-tale 'gefuffle' of the exhaust was missing, so the boat must have been running some item of noisy equipment such as a compressor. At 11.22 hrs *Venturer*'s first lieutenant sighted both the other submarine's periscopes (or possibly one periscope and the HF W/T mast, since Wolfram would have been anxious to make contact with his shore base to notify his presence and prevent attack by friendly forces). The masts were kept up for quite some time and their use was subsequently described by *Venturer*'s flotilla commander as 'most shamefully careless'.[1] From the information provided by Asdic and periscope observation, Launders concluded that he was broad on the target's starboard bow.

For the next hour Launders used his very primitive Type 129 Asdic to produce a plot of the target's course and at 12.12 hrs, when reasonably sure that the plot was correct, fired four Mk.VIII torpedoes in a 'hosepipe' salvo. The range on firing was reckoned to be 2,000yd but judging by the time it took before the first torpedo hit, it was probably more like 3,000yd. One explosion was heard on the right bearing after 2 min 12 sec running time. Two more explosions were heard after 5 min, indicating that the torpedoes had detonated on the seabed. Two minutes after the first explosion Asdic reported breaking-up noises and the ominous sound of rushing water. *U864*'s bulkheads were designed to survive a surface collision – they could not survive pressure at depth. Launders took *Venturer* to the scene to conduct an inspection through his periscope. There was an extensive slick of diesel, a considerable amount of wooden wreckage, probably from the U-boat's extensive casing, and a large metal canister which may have been one of the upper deck lockers used for

The British V-class submarine HMS Venturer. *On 9 February 1945 she sank U864 – both boats were submerged at the time and this attack remains the only time that one submarine has sunk another while both were submerged. (Author)*

carrying dinghies or life-jackets but may have been the container for a Fokker *Aghelis* tethered autogyro used for reconnaissance purposes.

The number of submarine engagements since 1945 has been extremely small with most of the action confined to the Indian Ocean where India and Pakistan have been engaged in a long-running conflict since 1947. On 4 December 1971 a Pakistani submarine attacked the Indian aircraft carrier *Vikrant* off East Pakistan (now Bangladesh) but the torpedo missed. The submarine was later, and incorrectly, claimed as sunk by India. On 9 December 1971 the Pakistani submarine *Hangor* (a French-built Daphne-class SSK) attacked the Indian frigate *Khukri* in the Arabian Sea. The commanding officer fired nine torpedoes (Daphne-class SSKs carry eight internal and four external tubes) and scored hits with three of them. *Khukri*, a small British-built Type 14 frigate, blew up and sank with the loss of 191 of her 288 crew.

In 1982 war[2] broke out in the South Atlantic between Britain and Argentina following the latter's invasion of the Falkland Islands. Both Britain and Argentina possessed a submarine force but at opposite ends of the *matériel* spectrum. Argentina possessed two ex-US Navy GUPPY conversions and two German-built Type 209 SSKs. Britain possessed a

HMS Conqueror, *which visited Portsmouth in October 1982.*

much larger submarine force consisting of SSBNs, SSNs and SSKs. Argentinian submarines achieved no successes, and their operations were characterised by *matériel* failure, lack of training, failure to press home attacks and fear of British ASW counter-measures. It was, however, a Royal Navy SSN that became the first nuclear submarine to fire a torpedo in anger and score a success.

The sinking of the Argentinian cruiser *General Belgrano* by HMS *Conqueror* (Cdr C. Wreford-Brown RN) on 2 May 1982 has been the subject of much comment, largely by those who saw the sinking as a foul British plot to derail a peace initiative being brokered by the Peruvian government. The decision to sink the *Belgrano* was a purely naval one, taken for tactical reasons and nothing else. As the British fleet made its way south it was menaced by a two-pronged Argentinian naval attack. To the west was a battle group centred around the aircraft carrier *Venticinco de Mayo*, while to the south was the cruiser *General Belgrano*[3] and her two escorts. The Argentinian plan was clear – aircraft would attack and damage the two British carriers after which *Belgrano* would arrive to finish off the cripples. Alternatively, *Belgrano* and her consorts would appear from the south just as the British were preparing to receive an air attack from the north-west. Either situation was not a comfortable one.[4]

Britain had three SSNs in the South Atlantic, *Spartan* (Cdr J. Taylor RN), *Splendid* (Cdr Roger Lane-Nott RN) and *Conqueror*. The first two were looking for the carrier but had been unable to locate her. *Conqueror*, however, had found *Belgrano* and her escorts and was shadowing them, pursuing the 'sprint and drift' technique of making a fast passage at depth and then slowing while coming up to periscope depth to take observations.

The British battle group commander, Rear Adm John Woodward, wanted to remove one of the two Argentinian 'pincers' threatening his force. Since he was not in contact with the carrier it would have to be *Belgrano*. She and her escorts were a strong group, operating well within range to interfere with any landings on the Falkland Islands and were believed, correctly, to be manoeuvring with other surface forces for a co-ordinated attack. Moreover, with a worst case appreciation that an Argentinian air attack might sink 25 per cent of British ships, no margin of strength existed for accepting surface ship-inflicted casualties as well. The next problem lay in that the operations of the British submarines were being controlled direct from London – against Woodward's advice. Thus he could not order *Conqueror* to attack since she was not under his command and such an order would require the ROE[5] to amend it – a political decision. Such a process would take time, a commodity that was in fairly short supply in the South Atlantic. Woodward took the decision to order the attack directly, knowing that London would see the signal, realise the urgency of his position and issue the necessary amendments to ROE. Which is what happened.

Conqueror received the amended signal at 17.30Z/2 and began to manoeuvre for the attack. The submarine was positioned some 7 miles astern of the cruiser and her escorts which were steaming in a V-formation, *Belgrano* to the south with one destroyer positioned half-a-mile off her starboard bow and the other the same distance off her starboard beam. The Argentinian ships were taking no precautions against attack: they were keeping a steady course and speed and making no use of the nearby shallow waters of the Burdwood Bank. *Conqueror* then dived deep and began a left-hand swing that would bring the submarine up on the cruiser's port side, just forward of the beam. Wreford-Brown had decided to use three unguided Mk.VIII torpedoes rather than the wire-guided Tigerfish in view of the latter's reputation for unreliability. At 18.57 hrs *Conqueror* came up on *Belgrano*'s port beam, almost on a perfect 90° angle to the Argentinian ship, at a range of 1,380yd, for final observations before the three torpedoes were fired.

Fifty-five seconds later the first torpedo smashed into *Belgrano*'s port bow, followed seconds later by another hitting further aft. *Belgrano*'s lack of external defence against submarine attack was mirrored by the lack of any internal precautions. The ship's company was not even closed up at defence stations and most of the ship's watertight doors were open – a fatal error in a warship at sea. All accounts speak of a fireball racing through the ship, indicating that standards of cleanliness between decks cannot have been very high. Damage control efforts failed to save the ship and she rolled over on to her port side and sank by the bow.

What of *Conqueror*? The submarine dived deep and moved away to the south-east after firing to avoid the destroyers. Sonar picked up the noise of a few desultory depth-charges being dropped. *Conqueror* observed the destroyers manoeuvring but did not attack. She then moved away leaving the destroyers to the unenviable task of rescuing the survivors. *Conqueror*'s action had the effect of persuading the Argentinian Navy to inflict sea denial on itself and for the rest of the war it remained in port. No better example of the usefulness of a submarine exists. What would Lt Paparrigopoulos have thought?

NOTES

1 Naval Staff History, *Submarines, Vol. 1, Operations in Home, Northern and Atlantic Waters* (London, 1953), p. 223.
2 Technically the Falklands conflict was not a war since neither side made a formal declaration of hostilities. However, the events following the Argentinian invasion of the Falklands represented nothing less than full-scale warfare.
3 Ex-USS *Phoenix*, armed with fifteen 6in guns and a number of Exocet surface-to-surface missiles.
4 See Woodward, Adm Sir John, *One Hundred Days, The Memoirs of the Falklands Battle Group Commander* (London, HarperCollins, 1992), pp. 146–55 for the most lucid exposition of the decision to sink *Belgrano*.
5 The ROE in force forbade attacks on Argentinian warships outside the Total Exclusion Zone around the Falkland Islands.

GUN ACTION!

The U-boat is inferior to every surface warship in a gun action.
German U-boat commanders' handbook, 1942

At first consideration the gun and the submarine are incompatible weapons. The submarine relies on the ability to dive both as a means of attack and as a means of escape. Gun action requires the submarine to be on the surface and highly visible. Moreover, whereas effective damage control can enable a surface ship to survive quite substantial gunfire damage, submarines are appallingly vulnerable to shell or splinter damage. One hit on the pressure hull and a submarine is transformed into a relatively slow and unmanoeuvrable surface craft. This view was accurately summed up in a handbook issued to U-boat commanders in 1942:

> The U-boat as an armament carrier is fundamentally a contradiction in itself. It is not constructed for gunnery action on account of its limited stability and its low and unsteady gun and observation platforms which are directly exposed to the sea's motion. Strictly speaking the U-boat is inferior to the surface vessel in a gun action. This is because the U-boat, as opposed to her surface opponent, is rendered completely vulnerable in a gun action since *one* hole in the pressure hull can prevent the U-boat from being able to dive and easily result in the loss of the boat.[1]

Moreover, submarines lack the sophisticated fire control instruments to enable a gun to be used to its full potential. They are low in the water and do not afford the control officer a position of any height from which he can direct the fire or observe the fall of shell. Fitting a gun armament to a submarine forced the designers, of whatever nationality, into a series of compromises. For a gun to be effective it must be fitted, along with the access hatch for the crew, as high as possible so that the gun can be brought into action in the quickest time. It also had to be sufficiently above the waterline so that the gun could be worked when the submarine was in the lowest buoyancy condition. At the same time considerations of stability and the size of the submarine's silhouette limited the height at which it could be placed. It may be physically possible to fit a large calibre weapon to a submarine, but the ammunition must not be so heavy or so large that it cannot be handled quickly by the loading party and the gun crew.

These disadvantages notwithstanding, the submarine and gun have forged an effective partnership in both world wars and afterwards.

Submarines have used guns both for sinking targets not worth a torpedo and for indulging in shore bombardment. In these roles the gun has been an effective addition to the submarine's torpedo armament. Also, but less successfully, submarines have used guns for defence, particularly against aircraft.

It is difficult to pinpoint which submarine was the very first to carry a gun. The US Navy's *Holland*, built in May 1897, was fitted with what was described as an 8in 'dynamite gun'.[2] The gun was worked by pneumatic power and could fire a projectile weighing 222lb carrying 100lb of guncotton to a range of a mile. Whether this gun was ever fired is not known but for a boat of 75 tons displacement and 53ft 10in long it is really doubtful if the weapon was practical. However, the first submarine to carry a gun in the conventional sense was the British *D.4* which was fitted with a 12-pdr gun on completion in November 1911. The weapon was mounted on a pneumatically operated cradle housed in a non-watertight casing in the conning tower. Although this fitting was an experiment (there was no room in *D.4*'s design for a magazine), there is no doubt that the British intended to fit submarines with a gun armament, most likely a 1½-pdr gun, from 1911 onwards. Designs for the first boats of the E-class show that space was allocated for a magazine. The intention was for a fixed mounting to be fitted to the casing: the gun itself would kept below and brought up to the casing when required.

The demands of active service in the First World War really brought the gun and the submarine together. Initially many British submarines carried light weapons such as the 2-pdr or the 12-pdr. There was no coherent scheme or policy behind the fitting of these weapons as submarines had to take whatever guns and mountings were available at a time when the Navy was undergoing tremendous expansion. The main purpose for these weapons was firing at the German Navy's Zeppelins which at that time enjoyed an undisputed monopoly of the air over the North Sea. The British submarine *E.4* carried no fewer than four 6-pdr guns on her casing for anti-Zeppelin duties – an unusual arrangement not fitted to any other submarine.

Operations in the Sea of Marmara gave British submarines the chance to use their guns offensively. Here British submariners found that their torpedo supplies rapidly became exhausted on account of the number of targets available. In *E.11* (Lt Cdr M.E. Nasmith RN) the torpedoes were set to float at the end of their run (in the event of a miss) so that they could be recovered and reloaded for the next attack. The British also found that a lot of the targets were so small (or of so shallow a draught) that they were simply not worth a torpedo. *E.11* was fitted with a 12-pdr gun for her second patrol in the Marmara (August–September 1915) and later exchanged this weapon for a 4in gun. *E.2* (Lt Cdr E. de B. Stocks) and *E.7* (Lt A.D. Cochrane RN) used their 12-pdr weapons for both sinking small vessels and shore bombardment. So successful did British submariners find the gun that *E.20* (Lt C.H. Warren RN) was fitted with a 6in howitzer specifically for shore bombardment purposes.[3] The campaign in the Sea of Marmara ended in January 1916 with the evacuation of British and Imperial forces from the Gallipoli peninsula. With the end of the campaign came, to all intents and purposes, the end of Britain's use of the gun as an

The 6in howitzer fitted to HMS E.20 for shore bombardment purposes. (Author)

offensive weapon. British submarines continued to carry guns, either the 12-pdr or the 3in, but their use would be almost entirely defensive, for use against German Zeppelins or seaplanes. They could be quite successful in this role – *E.31* (Lt Cdr Ferdinand Feilman) shot down a Zeppelin with her 12-pdr in 1917. Later British submarines of the L-class were armed with a 4in gun. This was considered the heaviest calibre weapon that could be conveniently worked by hand. There were no opportunities for gun action in the North Sea or Atlantic theatres and in the Adriatic, the only other theatre where British submarines operated, there is no record of any British submarine engaging in a gun action.

However, it was the Royal Navy which sent the largest gun to sea. The three M-class submarines were armed with a single 12in BL 30 cal Mk.XI gun taken from Majestic-class pre-Dreadnought battleships. The idea of fitting such a weapon was made clear in a statement by the First Sea Lord and RA(S) in August 1918:

The object behind the design was to supplement torpedo attack against surface ships which was found difficult against high-speed vessels steering a zig-zag course and which could often elude a torpedo when sighted by changing course. A 12-inch projectile fired at fairly close range should be difficult to elude. Furthermore fifty projectiles could be carried instead of a few torpedoes.

The boat and gun were designed so that the gun could be kept loaded submerged. A round already armed could be fired at a small target in thirty seconds from periscope depth, or twenty seconds after breaking surface. The vessel could be at periscope depth (24 ft) in fifteen seconds, i.e. before the smoke of discharge had cleared away. To reload it was necessary to surface but it was recognised that if the shot found the target the size was such that it would destroy the destroyer or light cruiser struck. In addition it could be used for bombardment.[4]

The gun and its director tower and loading chamber were on the forward side of, and incorporated into, the bridge structure, giving the submarines the appearance of a prehistoric monster. There was a common access compartment to the tower and the bridge. The 12in magazine with passageways either side, the 12in shell room and the hydraulic machinery compartment occupied about 30ft of the pressure hull directly beneath the gun. The weight of the gun and mounting was 120 tons and the 40 rounds of ammunition carried weighed a further 29 tons. The size of the mounting actually enhanced the submarine's diving qualities – the weight helped keep the boat down while its volume stabilised the boat at periscope-depth. The weapon was aimed from a director position forward of the bridge. A bead sight on the barrel was lined up with the target using a periscope-like instrument. The gun could be trained through 15° either side of the centre line, elevated 20° and depressed 5°.

However, the theory that the submarine could simply pop up and, in the words of one observer, 'Let him [the enemy] have a 850lb thump from nowhere when he was not expecting it' was severely compromised by practical limitations. The weapon was intended to be fired with only a portion of the barrel above the water, with aiming and spotting corrections being carried out from a periscope fitted in the director. In this scenario the gun would be at maximum elevation (implying maximum range) while the view from the periscope would be very limited – even more so if the sea state was anything greater than a flat calm. Only one of these submarines, *M.1*, was completed before the war ended. The Admiralty had been extremely concerned that the Germans would discover what was afoot and build their own version. The construction of *M.2–4* was delayed and then resumed although *M.4* was never completed owing to the end of hostilities. Plans existed for *M.1* to go to the Mediterranean and bombard either the Austrian base of Cattaro or the port of Constantinople but the war ended before these plans could be effected.

The German Navy took up the submarine gunnery business where the Royal Navy had left off. From February 1915 the Germans sought to bring Britain to collapse or at the very least to an armistice by destroying Britain's seaborne trade through a campaign of unrestricted submarine warfare. German submarine commanders found that it was far more effective to use their gun, either an 88mm, 105mm or in the case of the big U-cruisers a 150mm gun. British shipping, sailing independently, presented prime targets for gun action at a time when submarines did not have to consider the air threat. Undoubtedly the most successful German U-boat commander with the gun was KL Lothar von Arnauld de la Perrière of

The British submarine M.1 *with her characteristic 12in gun mounted forward of the bridge. (Author)*

U35. Perrière was not a submariner and brought a crack gun-layer from the High Seas Fleet to *U35* with him. During a patrol that lasted from 26 July to 20 August 1916 he fired over 900 rounds of 88mm ammunition and four torpedoes. (Exactly how that amount of ammunition was stowed on board the 685 ton submarine is not mentioned. Certainly *U35*'s practice cannot have been in accordance with any magazine regulations then in force!)

German U-boats' use of the gun depended on British shipping sailing independently and unescorted along known routes where the U-boats could lie in wait for them. Two factors began to restrict the use of the gun by U-boat commanders. The first was the employment of Special Service Vessels or Q-ships by the British. Converted to resemble the most humble of tramp steamers or sailing vessels, these warships packed a powerful punch in the shape of a concealed gun and torpedo armament. They patrolled the shipping lanes and hoped to entice a U-boat to the surface and within range of their armament. The disguise would then be dropped and the main armament would open fire. Actually Q-ships were far less effective than either side hoped or feared. Their first success was the sinking of *U36* (KL Ernst Graeff) on 24 July 1915 by the decoy *Prince Charles*. *U36* came to the surface to sink the 'steamer' and Graeff came within 600yd of the *Prince Charles* before she opened fire. Over eighty such vessels were commissioned in the Royal Navy but they accounted for

U35 *unloads her spent 15cm cartridge cases at Cattaro in the summer of 1917. The number of 'empties' in the lighter testifies to the extensive use made by de la Perriere of his gun. (U-boot Archiv)*

Commerce raiding – Handelskrieg – in the Mediterranean in 1917. U35 *shells the merchant ship* Parkgate *after her crew had abandoned ship. (U-boot Archiv)*

only eleven U-boats against a loss of twenty-seven of their number. Nevertheless, fear of being caught by such a vessel made U-boat commanders extremely wary of surfacing for a gun action. The second factor which inhibited the Germans' use of the gun was the gradual introduction of the convoy system from April 1917. Instead of waiting around the 'choke points' for shipping to come to them, U-boat commanders now had to attack a convoy with its escort of destroyers and other warships. Gun action was not possible under such circumstances.

The interwar period saw two more unusual gunnery developments in submarines. The first was Britain's *X.1*, built in 1925. Quite why Britain built this submarine is unclear. At the time of her commissioning in September 1925 she was the largest submarine in the world and with her armament of four 5.2in guns in two twin mountings she had the armament of a large destroyer. She was designed as a cruiser submarine but against what would she operate? Britain had no potential enemy with a vulnerable merchant fleet against which she could prey and in any surface engagement *X.1* was horribly vulnerable to a single shell or splinter hit on her pressure hull.

The twin 5.2in mountings were sited fore and aft of the bridge. Beneath each mounting a circular gun trunk of about 4½ft in diameter ran from the mounting to the magazine situated in the lower half of the pressure hull. A 10ft diameter working chamber surrounded the gun trunk between the top of the pressure hull and the mounting. Each magazine carried 100 rounds of ammunition. Fire control arrangements for these weapons took up most of the bridge space. From the pressure hull a hatch led up to an upper control-room on top of which was the director tower extending to about bridge canopy height. The cover of the director tower could be raised vertically by 2ft and when shut was obviously watertight to full diving depth. A second hatch from inside the pressure hull led to a range-finder room alongside the upper control-room. A 9ft range-finder was fitted on the bridge aft of the director tower. The conning tower was abaft these gunnery spaces. *X.1* had a chequered career. She was a good deal more successful in service than the claims of her detractors might suggest. However, her success only gave more ammunition to critics of the design who argued that such a submarine was of more use to a country like Germany or Japan (or even the old enemy, France) who could use such a boat to deadly effect against Britain's trade. Positive reports about the boat from sea were suppressed[5] and she was eventually abandoned to a mooring in Portsmouth harbour.

Across the Channel France was building a submarine in the same style as *X.1*, although whether the *Marine Nationale* was influenced by the British design or whether they were just exercising Gallic flair is not known. The submarine *Surcouf* launched in 1929 carried not only two 203mm guns but also a small Besson MB411 floatplane housed in a hangar aft of the conning tower.[6] *Surcouf* was designed to support and reinforce France's extensive colonial empire in Africa and the Far East as well as engaging in the traditional *guerre de corse* so beloved of French strategists since the mid-nineteenth century. Her gun armament was carried in a twin mounting forward of the conning tower. The guns had an elevation of 30° and could train 130° either side of the centre line. They

The French submarine Surcouf – *note her twin 8in mounting forward of the conning tower. This photograph was taken on the Clyde in 1940 when, though French-crewed, she was under British operational control, hence the British-style pendant number.* (ECPA)

had a range of 27,500m. Directly aft of the turret but forward of the bridge was the director, equipped with a 3m range-finder.

There are more rumours concerning the short and unlucky career of *Surcouf* than about any other submarine, but lack of space precludes their discussion here.[7]

The gun became a standard addition to nearly all submarines built during the interwar period. The Royal Navy used the 4in gun for all classes except the S-class built during the late 1930s. These boats were fitted with the smaller 3in weapon, as were the U-class which were originally designed without a gun, but were subsequently fitted to carry one in the light of war experience. S- and U-class submarines carried their guns on a simple pedestal mounting on the casing forward of the bridge, with safety rails fitted to give the gun's crew some safety in rough weather. All other British submarines carried their guns in a breastwork mounting which formed part of the bridge structure. This gave much greater protection from the sea, while a splinter shield around the gun gave protection from enemy fire. Submarine gunnery in the Royal Navy was almost a religion, particularly on the China Station where success in the annual flotilla gunnery competition (in which a boat had to surface from

Opposite: *The forward twin 5.5in mounting of the British submarine* X.1. *(Author)*

HMS Taurus's *4in gun trained on the starboard bow. The breastwork mounting gave some protection from the sea. (Author)*

periscope depth, fire a prescribed number of rounds at a target and return to periscope depth within a given time) was taken as an indicator of a CO's proficiency.

However, as the war developed submarines began to request additional firepower in the way of light automatic weapons for AA defence and for sinking small coastal craft. British submarines carried the .303 Lewis or .303 Vickers machine-gun or the .303 Bren gun, if the Army could be persuaded to part with the latter. These weapons were simply carried on to the bridge and bolted to a pedestal mounting when required. HMS *Terrapin* was unusual in that she carried two .5in Browning machine-guns purloined from the Americans (and which must have been hideously awkward to manhandle up and down the conning tower). The weapon had immense stopping power but proved too heavy for the relatively 'soft' bridge structure, and the Brownings had to go.

The favoured weapon was the 20mm Oerlikon which proved extremely valuable, though not as useful in the AA role as had been expected. The 20mm Oerlikon was fitted to all British submarines except those of the U-class which lacked sufficient stability to compensate for the extra topweight. It was fitted into a bandstand at the after end of the conning

The 3in gun fitted to most British S- and U-class submarines during the Second World War fired a 17.5lb shell to a range of over 12,000yd. (Author)

tower. The bandstand was fitted with a safety rail to prevent the over-enthusiastic gunner from damaging the bridge and periscope standards. The weapon had a crew of two: a gunner and a loader who supplied the ammunition contained in pans, each holding 120 rounds. The gunner stood in a steeped well in the bandstand which allowed him to increase the weapon's elevation by stepping down into the well. The Oerlikon had considerable deterrent value but its effectiveness was marred by its tendency to jam and the cartwheel sight was less than effective against fast-moving targets. However, when used against junks and other native craft in the Far East it proved very useful. On the whole it was very well liked. HMS *Tactician*'s Oerlikon gunner compared his weapon to a woman: 'Plenty of attention, no neglect and she would not let you down.'[8]

During the Second World War the Royal Navy demonstrated a highly individualistic method of submarine gunnery. British submarines operating in the Mediterranean and the Far East encountered numerous targets all of which were too small to be worth a torpedo. However, British submarines did not enjoy the virtual immunity from anti-submarine measures which de la Perrière *et al.* had enjoyed in the First World War. British operations were conducted in waters close to land and were frequently patrolled by

The 20mm Oerlikon gun – a standard close-range weapon in British and American submarines during the Second World War. (Author)

enemy forces. Gun action therefore had to be effective and above all quick. No British submarine commander could afford to hang around on the surface while his gun-layer engaged in lazy target practice.

British submarines perfected the drill of not surfacing until the very last moment in order to achieve complete surprise. On sighting a suitable target through the periscope the crew were called to diving stations. In T- and S-class submarines the gun crew had their own access hatch to the gun platform forward of the conning tower. However, in the U-class (which bore the brunt of operations in the Mediterranean) the gun crew had to race up the conning tower ladder to the bridge and then fling themselves over the forward face of the conning tower down on to the casing. Meanwhile the gun-layer and trainer would be given an opportunity to inspect their target through the periscope while the captain made the usual observations of range, angle on the bow, and speed. If the submarine possessed proper gunnery control instruments then these observations would be set on the control-room transmitter. If no such instruments were available then the data would be given to the sight-setter to put on the gun when he reached his position. The submarine would also be manoeuvred into a suitable attacking position – 1,000–2,000yd abaft the target's beam was considered a good position.

The captain then gave the order 'Salvoes shoot', followed by 'Man the gun tower'. The gun crew would gather either in the gun tower or, in the case of U-class submarines, in the conning tower. The order 'Stand by to surface' was then given, after which the gun-layer removed the pins that secured the upper hatch. Only water pressure was keeping the hatch shut. The captain now had to bring the boat to the surface in the quickest fashion possible. Accordingly the hydroplanes were put to dive while speed was increased to give the hydroplanes more effect: these actions effectively kept the boat down. At the same time the first lieutenant bled HP into the boat until three or four inches above atmospheric pressure showed on the barometer. The order 'Surface' was then given at which all ballast tanks were blown. Two opposite forces were now working on the submarine: the motors and hydroplanes were holding her under while the blowing ballast tanks and positive pressure in the boat were forcing her up. Eventually it was the upwards pressure that prevailed and the submarine would rise to the surface extremely quickly. When the depth-gauge passed the 20ft mark, the first lieutenant (now the senior officer in the control-room since the commanding officer was effectively incommunicado in the conning tower) blew a whistle. While the planesman put the hydroplanes to 'hard to rise' (thus removing the last constraint holding the boat down), the gun-layer forced the hatch open. Although the hatch was still under water the excess pressure in the boat helped open the hatch and the gun's crew raced out.

The first round was fired for effect. The gun would be cold (a factor that affected the range) and the crew would not have settled at their positions. Nevertheless, the shell, however inaccurate, would have a demoralising effect on the enemy. A good gun crew could achieve a rate of fire of ten rounds per minute and it was not long before the target was out of action or sinking.

While the gun action was going on the look-outs on the bridge had to maintain their vigilance despite the distraction of events unfolding around them. A submarine could not afford to be surprised on the surface by an aircraft. If an aircraft did appear then the gun crew, look-outs, signalman, OOW and commanding officer had to get back into the boat as quickly as possible, regardless of crushed fingers or the state the gun was left in. During a hurried dive in HMS *Trespasser* in 1942 no fewer than seven men (and a Bren gun) were crammed into the conning tower between the upper and lower hatches!

Gun action was a totally different sort of warfare for British submariners. By way of contrast with a torpedo attack it was noisy, involved a large portion of the ship's company and offered an instant success that could be discussed at leisure later in the fore-ends. Moreover, since nearly all targets attacked with the gun were unescorted merchant vessels, gun actions carried little risk of retaliation from the escort. In the words of one author, 'Gunnery was, frankly, a good game.'[9] However, not all British gun actions took place against small, undefended targets. On 4 March 1945 the submarines *Trenchant* (Cdr A.R. Hezlet RN) and *Terrapin* (Lt R.H. Brunner RN) attacked a Japanese convoy in the Malacca Straits and sank submarine chaser *No. 5* and a coaster while forcing three other coasters to beach themselves before destroying them with gunfire.

The first German U-boats built after the First World War carried a single gun, either a 12cm in the case of the Type 1A, or a 20mm in the

Gun action in the Lombok Strait. Cdr Arthur Hezlet watches as shells from HMS Trenchant *fall around the Japanese Special Minesweeper No. 105 on 24 May 1945. (Jim Gilbert)*

case of the Type IIs, on a pedestal mounting forward of the conning tower. The Type VII series of boats which were the backbone of the German submarine fleet carried an 88mm gun forward of the conning tower and a 20mm gun at the after end of the conning tower or sometimes on the after casing. Larger Type IX and X boats carried a 105mm gun.

German U-boats used their deck guns less frequently for attacks than British or American submarines. U-boats operated in the main as submersible torpedo craft against defended convoys where there was little scope for gun action, other than finishing off a cripple damaged in a night convoy action. Only in waters far from patrols such as the Caribbean or Indian Ocean or in actions against independently routed ships would the U-boats use their deck guns to any degree. However, following the defeat of the Wolfpack tactics in the Atlantic in May 1943 and the appearance of both long-range aircraft and 'hunter-killer' escort groups in the Atlantic, the gun assumed a new importance to the U-boat as a means of defence against aircraft. Four-barrelled 20mm AA guns were fitted to protect U-boats against air attacks, particularly when transiting the dangerous waters of the Bay of Biscay. The AA armament was gradually increased and was mounted in two platforms at the after end of the conning tower.

The lower platform held either a single 37mm AA gun or a quadruple 20mm mounting. The upper platform carried two twin 20mm mountings. Six Type VII U-boats (*U211*, *U256*, *U271*, *U441*, *U621* and *U953* were converted to anti-aircraft boats and were armed with one 37mm, a quadruple 20mm mounting and two twin 20mm mountings. This constituted formidable firepower and the BdU felt that the increased armament and the policy of sending the boats out on patrol (and returning from patrol) across the Bay of Biscay in groups so that their firepower would be mutually supporting would prove an effective counter to Allied air attacks. Early encouragement was given to these 'fight-back' tactics when *U758* successfully defended herself against attacks by VC-9 aircraft embarked in USS *Card*.

The BdU's hopes were not to be realised. Submarines are always at a disadvantage when defending themselves against aircraft. One aircraft could be driven off or shot down by a submarine but Allied aircraft soon learned to circle and call up reinforcements, either more aircraft or surface ships. One submarine could not take on several aircraft, and U-boat commanders were faced with a difficult choice when confronted with aircraft. Should they dive and hope to escape into the depths unscathed? This would mean diving and laying the undefended boat open to an attack at the moment when all personnel were below (and the guns unmanned) and when the submarine was lying awash. Or should they hope to escape on the surface? Sooner or later the aircraft would reach their prudent limit of endurance (PLE) and have to go home – but doubtless the aircraft would have broadcast the submarine's whereabouts to any available reinforcements. It was a hard choice that was not made any easier when aircraft, particularly carrier-borne aircraft, began operating in groups, with some configured for dealing with a U-boat that chose to stay on the surface and fight it out while others carried depth-charges and acoustic torpedoes.

On 9 August 1943 *U664* (KL Adolf Graef) was spotted by a TBM Avenger flown by Lt(jg) C.G. Hogan USNR from USS *Card*. Instead of attacking immediately Hogan called up a team from the *Card* which consisted of an F4F armed with two impact fused bombs and two TMB Avengers armed with depth-charges and mines. Lt Hodson flying the F4F was first to attack with a strafing run, followed by Hogan with two 500lb bombs. Graef now decided to take *U664* under but as the boat was diving Lt(jg) Forney roared in and dropped two depth-charges. *U664* was blown back to the surface by the explosions. Inside the U-boat there was pandemonium. The *LI* requested permission to abandon ship but his question was treated as an order by about fourteen of the crew who rushed up on to the casing where they were killed by Hodson's second strafing pass. Graef now decided to dive for a second time but at a depth of 50ft it was clear that the hull was leaking so badly he had no alternative but to surface and surrender. As his crew went over the side *U664*'s bow rose higher and higher out of the water and at 14.20 hrs she sank stern first. The survivors were rescued by the USS *Borie* which had to break off operations when attacked by another U-boat.

Even in groups, U-boats were not safe. On 30 July the Second Support Group (Senior Officer, Cdr F.J. Walker RN) was hunting in the Bay of

Biscay supported by an RAF Sunderland and a Catalina. At 07.14 hrs *Wild Goose* picked up an HF/DF bearing on a U-boat sending a long signal and the Group, now joined by an American Liberator and two Halifax, closed the position at speed. At 10.05 hrs three conning towers were visible and Cdr Walker hoisted the signal 'General Chase'. The three U-boats were in line abreast with *U461* flanked by *U462* and *U504*. Their combined firepower had already beaten off the first air attacks but the standing Luftwaffe patrol of nine Ju.88s had insufficient fuel to reach them.

The aircraft returned for a second attack and a Halifax of 502 Squadron damaged *U462* so that she could not dive. Then the Liberator went in low (so low that it drew the concentrated fire of all three boats) was badly shot up and had to head for the safety of Portugal following the attack. Taking advantage of the diversion caused by this attack, the Sunderland came in from astern and sank *U461* which disappeared very quickly. By this time the ships of the Support Group were within range and they opened fire on *U462* (ObLtz.s Bruno Vowe), firing a total of 112 rounds. HMS *Kite* registered a hit at the remarkable range of 13,050yd and was rewarded by a message from a Halifax viewing proceedings that the U-boat was sinking. Vowe saw the sloops coming up and with his boat damaged by gunfire, decided to scuttle the submarine.

U504 had dived when the shells from the ships started to fall around her. *Kite* was quickly in contact but Walker took his time, allowing *U504* to move away from the large number of survivors from *U461* and *U462* who were in dinghies on the surface. The first two attacks were unsuccessful and it was clear that *U504* had gone deep. A series of creeping attacks were launched with the *coup de grâce* being given by HMS *Woodpecker* and *Wild Goose*. The echo on *Kite*'s Asdic then faded and the usual grim evidence of a kill came to the surface. Three U-boats sank – an excellent example of air–ship co-operation. The impressive AA armament fitted to U-boats could not compensate for the submarine's inherent disadvantage when operating against aircraft.

American submarines generally carried a 3in gun on a pedestal mounting placed either forward or aft of the conning tower. US submariners were not such keen exponents of the gun action as the British and when they did so preferred to stalk their prey on the surface, using their higher surface speed and better radar, and would manoeuvre so as to use the light to the best advantage. However, the development of submarine gunnery in the US Navy during the Second World War mirrored British practice in that submarine commanders wanted more gun armament, both for self defence and for sinking targets not worth a torpedo. US submarines had their bridges cut down to enable more guns, 20mm and 40mm weapons, to be mounted. The 3in deck gun was replaced by the 4in and ultimately the 5in gun. *SS285* to *SS291* of the *Balao*, *Gato* and *Tench* group were armed with one 4in/50, one 40mm and two 0.5in machine-guns. *SS292* to *SS312* had one 5in/25, one 40mm and two 0.5in machine-guns. *SS313–352, 365–378, 381–426, 435, 475–490* and *522–525* were armed with one 5in/25, one 40mm and one 20mm. The exceptions to this pattern of gun armament were the three large submarines *Argonaut*, *Nautilus* and *Narwhal*, each of which carried two 6in/53 guns fore and aft of the conning tower.

Japan's extensive submarine fleet carried either a 4.7in or a 5.5in gun as well as a variety of smaller close-range weapons. Japanese practice in this respect follows the generally lacklustre use of their submarines. Occasionally a Japanese submarine would sink an Allied vessel by gunfire but that remained the extent of Japanese submarine gunnery. Other nations' efforts in this area can be quickly dealt with. The large French submarine fleet was quickly removed from hostilities by the Armistice of July 1940. French submarines carried a 3in/35 Mod 28 deck-gun supplemented by either a twin or two single 13.2mm guns carried on the bridge. Italian boats carried either a single or twin 100mm or 120mm deck-guns but Italian submarines also carried a large number of 13.2mm machine-guns which could be quickly brought up on to the bridge or casing.

In the postwar period the gun disappeared from submarine use with incredible speed. There are two reasons for this. First, the change in submarine warfare meant that streamlining and submerged speed were now factors of major importance. The gun constituted an important and significant drag factor. Submarine warfare was also changing: the main target was now other submarines in engagements where the gun would be of limited use. Secondly, as the surviving wartime boats were fitted with the schnorkel underwater breathing tube, the gun was removed in order to save top weight. However, in British service at least, the gun enjoyed a brief renaissance in the early 1960s during the confrontation with Indonesia. A number of SSKs of the Oberon-class carried a 20mm gun on the casing while several submarines of the streamlined A-class carried a 4in Mk.XXIII gun forward of the fin. The end of submarine gunnery in the Royal Navy came in December 1974 when HMS *Andrew* fired the final round marking the end of an era with the following signal: '*The reek of cordite has passed from the Royal Navy's Submarine Service. Last gun action conducted at 031330Z. Time to 1st round – 36 seconds. May the art of submarine gunnery rest in peace and never be forgotten.*'[10]

In one respect, however, the AA role of the gun may well be taken over by the guided missile. All large navies now possess helicopters configured for ASW operations and a pair of these machines, fitted with dunking sonars and homing torpedoes, can prove tough opposition. A decoy may be sufficient to frustrate a single attack but not a second. However, the ability to shoot down at least one helicopter would increase the chances of escaping from the other. In the early 1970s the Royal Navy experimented with SLAM (Submarine-Launched Airflight Missile), a short-range anti-helicopter missile system. The weapon consisted of a retractable mast mounted at the forward edge of the conning tower of HMS *Aeneas*, containing four Blowpipe wire-guided missile launchers grouped round a TV camera. A monitor in the control-room allowed the commanding officer to identify a target and then aim the missile. The system worked but had the disadvantage of revealing the submarine's position the moment the mast was raised above the water. After lengthy trials the Navy decided not to accept the system. Given the increased importance of the helicopter in ASW in the twenty years since SLAM, it can only be a matter of time before these missiles make their appearance in the submarine armoury.

HMS Aeneas *showing the SLAM launcher fitted to the forward end of her conning tower. The four missile canisters can be seen grouped around the central TV camera. The development of anti-helicopter SAMs like SLAM is one certain progression in the continuing story of submarine armament.*

Another factor working towards the introduction of some gun armament in submarines is the changed nature of operations in the post-Cold War world. Flexibility and littoral operations are the new watchwords for naval planners and financial bean-counters alike. Submarines have to justify their place in an order of battle which is increasingly dominated by financial constraints. Therefore, they must be seen to be as flexible as their surface counterparts. The provision of a gun on some kind of retractable mounting that folds back into casing or fin (to satisfy streamlining and acoustic housekeeping requirements) should not be beyond the wit of submarine designers. The provision of a gun would allow submarines to play an effective role in, for example, sanctions enforcement operations. A submarine would be less vulnerable to attack by FAC, shore-based missile or suicide craft than a frigate and the opposition would never know where the submarine was. The reek of cordite may return.

In concluding this chapter it is worth noting that only one submarine, HMS *Triad*, has been sunk in a gun action with another submarine. On 15 October 1940 *Triad* (Lt Cdr J. Salt RN) was on patrol in the Gulf of

The British submarine Swordfish *showing her 3in gun positioned on a 'disappearing' mounting which could be recessed into the casing. (Author)*

CC Bandino Bandini relaxes in the control-room of his submarine Enrico Toti. *On the night of 14/15 October 1940* Toti *sank the British submarine* Triad *in a surface gun and torpedo action. This is the only occasion when one submarine has sunk another in this fashion. (Dott Achille Rastelli)*

Taranto in the Mediterranean and encountered the Italian submarine *Enrico Toti* (CF Bandino Bandini) on the surface shortly after midnight. There was a brief but intense gun duel in which *Triad* scored two hits on *Toti*'s casing, causing little damage, and then fired a torpedo which passed astern of the Italian boat. Both submarines now passed each other so closely that those on *Toti*'s bridge could clearly hear *Triad*'s bridge party speaking English. A sailor on *Toti*'s casing threw his boot at the British boat! *Toti* now used her superiority in close-range weapons, 'a real hail of fire'[11] according to Bandini, to force *Triad*'s bridge-party and gun crew below – a tactic that succeeded. *Toti* then fired a torpedo but the range was too close and the weapon had not armed by the time it struck. *Triad* began to fire but as she did so *Toti* scored two hits on her pressure hull near the conning tower which ignited the recognition flares kept on the bridge. *Triad* was then seen trying to surface stern first but she eventually sank by the bows.

NOTES

1 *Handbuch fur U-bootskommandanten* (1942).
2 Fyfe, Herbert C., *Submarine Warfare* (1902), p. 54.
3 *E.20* never used her 6in gun in action. On 5 November 1915, less than three months after commissioning, she was sunk by the German *UB.14*.
4 MoD (Ship Department), *The Development of HM Submarines*, BR 3043, January 1979, p. 9.1.
5 The late Capt Gilbert Roberts to author, 7 August 1984. Roberts was *X.1*'s gunnery officer.
6 For full particulars on *Surcouf* see Claude Huan's excellent *Les Sous Marins Français, 1918–1945* (Marines Editions, 1995).
7 See Compton-Hall, Richard, *Monsters and Midgets* (Blandford, 1985); and Rusbridger, James, *Who Sank Surcouf?* for a fuller examination of this submarine's story.
8 Gus Britton to author, 28 January 1987.
9 Compton-Hall, Richard, *The Underwater War* (Poole, Blandford Press, 1982), p. 65.
10 RNSM Archives.
11 Ufficio Storico della Marina Militare: La Marina Italiana Nella Seconda Guerra Mondiale, Vol. XIII, I *Sommergibile in Mediterraneo, Dal 10 Guigno 1940 al 31 Decembre 1941* (Rome, 1972), p. 96.

WEAPONS THAT WAIT

The mine and submarine are an ideal marriage of weapon and delivery system. The mine is the perfect weapon of position which can be left at the entrance to an enemy's harbour or across a known enemy transit route or choke point. Even before the First World War the mine had exercised considerable influence on events in the Russo-Japanese War and the lessons of that conflict were not lost on naval staffs planning for a European War. Up to 1915 the biggest problem was always in delivering the mine to the right position and then making a speedy exit. Surface minelayers were vulnerable to detection. The first German naval casualty of the First World War was the minelayer *Konigen Luise* which was caught and sunk by cruisers and destroyers of the Harwich Force on 5 August 1914.

The submarine offered the ideal delivery system. Providing that a safe system of laying the mines could be developed then the minelay could be conducted without the enemy being aware that anything had happened until the first ships started to sink. It was the Russians who built the first dedicated submarine minelayer, *Krab*, possibly as a result of their shattering experiences at the hands of the Japanese during the Russo-Japanese War. Commissioned in the summer of 1915 but not begun until 1908, *Krab* was the brainchild of a Russian railway engineer called Naletov who had never designed a ship before let alone a submarine. She was nicknamed the 'Box of Surprises', on account of her temperamental behaviour – not least the fact that she took over 20 minutes to dive! After completion, the crew were left to discover for themselves how the boat's various systems worked. Sixty mines were carried on two parallel tracks inside the casing on top of the pressure hull. They were launched over the stern by means of an endless chain conveyor belt. *Krab* was employed exclusively in the Black Sea and laid a number of fields but only claimed one victim, the Turkish gunboat *Issa Reis*.

It was the Germans who made the first successful use of the dedicated submarine minelayer. In support of the first campaign of unrestricted submarine warfare, the Germans proposed employing submarine minelayers around the British coast. These were the small boats of the UCI-type, *UC1-15*, which carried twelve UC120-type contact mines in six free-flooding chutes located in the forward section of the pressure hull. No torpedo armament was carried. After being released a soluble plug was supposed to keep the mine safely tethered to its sinker until the submarine had passed over. The mine would then rise to the surface. However, some plugs dissolved more quickly than others, and five UC boats were destroyed by their own mines rising up underneath them. The most notable victim was *UC12* (ObLt.zS. E. Frohner) which was blown up on her own mines on 16 March 1915 outside the harbour of Taranto in Italy. Germany was not even formally at war with Italy at the time – an example of how a submarine can be used to circumscribe diplomatic convention – but *UC12* had been 'transferred' to the Austrian Navy and carried an Austrian liaison officer even though she retained her German crew. When the Italians salvaged the boat and discovered that she was laying mines they were extremely annoyed at Germany's actions. The fact that the submarine was German was a major influence in Italy's joining the *entente powers*.

German minelaying submarines were initially employed in the North Sea, the Thames Estuary and the English Channel. These waters lay within easy reach of the newly acquired bases at Zeebrugge and Ostende and meant that the submarines did not spend too much time on passage. Accurate navigation was essential for minelaying, so that one's own forces knew where the danger areas were. Accordingly, German submarines were fitted with a sophisticated sounding device worked from within the boat to make navigation as accurate as possible. The UCs were fairly successful but their usefulness declined as British ASW measures – counterminelaying and net barrages – began to bite in the English Channel. Nevertheless, the Germans went on to build another seventy such UC boats with later variants, the UCII and III types also incorporating a torpedo armament. The Germans also constructed eighteen larger minelayers, eight of the UE type (36 mines carried) and ten of the U117 type (48 mines carried). The most promising hunting-ground for Germany's submarine minelayers lay off the eastern seaboard of the United States following America's declaration of war in April 1917 but only a small effort was made to mine these ports. Only fifty-eight mines were laid, yet they accounted for four merchant ships sunk along with the cruiser USS *San Diego*. The battleship *Minnesota* was also damaged by a mine. More importantly, several major ports were closed until hastily improvised sweeping operations could be implemented. This last action threw the transatlantic supply route into total chaos.

The German inability to drive home a minelaying campaign in American waters as they had done in UK waters in 1915 was due to the fact that the *UE*s and the U117-class lacked the range to reach the east coast of the United States from Europe. By the time the enhanced endurance became a requirement, German shipyards were so overloaded with work that the necessary conversions could not be accomplished. Germany had provoked the United States into declaring war through her third campaign of unrestricted submarine warfare. Having done so, the Germans neglected the one weapon in their armoury which they could use to strike directly at the United States.

However, the *UE*s were responsible for the sinking of some major British ships in home waters and in the Mediterranean. *U75* laid a minefield off the Orkneys, ironically in the wrong place owing to inaccurate navigation, which claimed the cruiser *Hampshire* on 5 June 1916. *Hampshire* was *en route* to Russia with Lord Kitchener, and his

Mines being loaded onboard the minelayer E.46 *at Brindisi in 1918. (Author)*

staff as passengers. In the Mediterranean *U73* (KL Gustav Siess) laid mines which claimed the battleship *Russell* and two escorts. Siess's mines also accounted for the 12,009-ton liner *Burdiglia* on 14 November 1916 and the 48,158-ton *Britannic* (sister-ship to the ill-fated *Titanic*) on 21 November. The *Britannic* was the largest single merchant ship casualty of the First World War.

Britain gained a fair amount of intelligence on submarine minelaying through the examination and salvage of sunken UC boats around the British Isles. In October 1915 orders were given to fit six British E-class submarines (*E.24, E.34, E.41, E.45, E.46* and *E.51*) with mining chutes in their saddle tanks in place of the beam torpedo tubes. Five mine chutes were built into the saddle tanks, two mines being carried in each tube, making a total cargo of twenty mines. On 4 March 1916 *E.24* (Lt Cdr George Naper RN) sailed on the first minelaying patrol by a British submarine, carrying twenty mines to be laid off the Elbe estuary. Even before this operation minelaying was unpopular. The mines had to be laid in shallow waters where navigation was difficult. Worse still, the minelayers had orders not to attack enemy shipping for fear of compromising their position and the minelay. Lastly, unlike a torpedo attack, there was no visible reward for their efforts in the shape of a sinking. Not the least of the submariners' many dislikes was the task of checking the mine wells after an operation, to make sure that all the mines had released and that none had 'hung up' inside the chute. This was done by the simple, but potentially lethal, process of pushing a pole down inside the chute to see if anything was there! (The reader should bear in mind that these were contact mines fired by breaking a contact 'horn' fitted on the upper end of the mine.)

The crew of the minelayer E.34 taken shortly before the submarine was sunk in July 1918. The mine wells in the submarine's saddle tanks are clearly visible on the port side of the submarine. (US Navy)

E.24's first operation was nothing short of a nightmare. The twenty mines bumped and rattled audibly during the passage and did nothing for the crew's nerves. Naper had no sounding device, like that fitted in the UC boats, so he had to find the correct patch of 15m deep water at the mouth of the Elbe by dead reckoning, notoriously inaccurate, and by taking periscope observations. Lt Charles Allen, the first lieutenant, commented that E.24 navigated 'largely by bouncing off the shallows, charting the depth all the time'.[1] E.24's field was probably laid miles from the correct position and thus constituted a hazard to both sides. The consequences of flawed navigation were serious. In July 1918 E.34 (Lt R. Pulleyne RN) almost certainly strayed into a field laid by E.51 off Vlieland on 3 April and was sunk. E.51's mines had been fitted with sinking plugs that would cause the mine to sink thirty-eight days after being laid so that the mine would not constitute a permanent hazard to operations. In this case it is clear that one of those sinking plugs failed to function.[2]

France and Austria did not commission minelayers, which was perhaps to their advantage since the losses suffered by these boats during the First World War were immense. Two British boats were sunk, E.24 and E.34, while the Germans lost no fewer than fifty-two minelayers, either through the explosion of their own mines, as in the case of *UC12*, or while engaged in minelaying operations.

During the interwar period specialist minelayers were built by nearly all the major navies with the exception of Japan's. Generally these were dedicated minelayers which laid their mines through wells or chutes in the saddle tanks, such as the French *Rubis* and German Type XA and XB boats. Alternatively, some submarines used a system like that of the Russian *Krab*. The British Porpoise-class used an almost identical system with one track of mines running the full length of the pressure hull underneath the casing. The mines were pulled to the stern by means of an endless chain conveyor. The system had been tested in *M.3*, the last of the three M-class submarines which had been armed with a 12in gun, and found to be successful. In order to let the mine casing run the full length of the hull, the periscopes and other masts were offset to starboard instead of being placed on the centre line as was the usual practice.

Dedicated submarine minelayers like the British Porpoise-class soon found themselves being 'borrowed' for other tasks. Store carrying was a particular favourite. The submarines' high mine casing made an excellent 'cargo compartment'. The American minelayer *Argonaut* was hastily converted in 1942 to a submarine transport, her mining facilities removed and replaced by accommodation spaces for 120 marines.

The German Type XA and XB boats laid 327 mines off the American east coast during the Second World War. Eleven ships were sunk or damaged. This was not a large return (one ship for every 29.8 mines) but seven ports were closed for forty days and, as in the First World War, the transatlantic supply route was disrupted. More importantly, the Americans laid several anti U-boat minefields which claimed a number of Allied ships lost or damaged.

The introduction of a mine that could be laid through a standard torpedo tube meant the end for the dedicated minelayers. These tube-laid mines could be either contact-fused or fitted with magnetic or acoustic firing circuits. Submarines could now carry a mixed load of mines and torpedoes. German submarine minelaying off the east coast of England claimed the brand new cruiser HMS *Belfast* on 10 December 1939. Though the ship returned to port, she was under repair for over three years. In the Pacific, Allied submarines laid 658 mines, nearly all by standard submarines using tube-laid mines.

No mention of minelaying in the Second World War is complete without reference to the French submarine *Rubis* and her incomparable canine mascot Bacchus. Built in 1933 *Rubis* (and her five sisters) employed the Normand-Fenaux system of minelaying which was based on that of the British E-class in the First World War. There were eight wells on either side of the submarine, each of which held two mines. *Rubis* was working under Royal Navy operational control in Britain when France signed the Armistice in July 1940. After a period of indecision, *Rubis*'s ship's company elected to remain in Britain and continue the fight with General de Gaulle's Free French forces. However, her Royal Navy liaison officer noted that, '*Rubis* is a French submarine, not a Free French one . . . patriotism and loyalty is not to de Gaulle but to France and is as strong as an Englishman's to England . . . If they are captured by Germans they will

HMS Narwhal loading Mk.XVI mines at Immingham in 1940. One mine is shown on the casing while another is positioned at the end of the mine casing, ready to be trundled up the submarine. (Author)

HMS Tetrarch, *the last dedicated British-built submarine minelayer, at Alexandria in 1941. (Author)*

not be shot but if they are captured by Vichy Frenchmen they will certainly be shot. They are fighting Germany for France, not for Britain.'[3]

Rubis was a spectacularly successful minelayer. In twenty-eight patrols she laid 683 mines which resulted in the sinking of fourteen supply ships totalling 21,410 tons, seven A/S ships and damage to a U-boat and a supply ship.

Undoubtedly her 'finest hour' came when returning from a minelay off Norway in August 1941. During the patrol she had been damaged by the shock wave from the explosion of two of her 18in torpedoes when they hit a supply ship, and she was consequently unable to dive. She was only 40 miles from the Norwegian coast. A battery fire below brought the entire equipage on to the casing. The cruiser *Curacao* and the destroyers *Lively*, *Lightning* and *Wolfhound*, together with a rescue tug, were sent to screen her. While waiting for rescue CC Cabanier, the commanding officer, ordered luncheon to be served using the galley in the conning tower. Soup was followed by cold ham and salad and an assortment of cold and boiled vegetables. Fruit salad and dried apricots constituted the dessert while sherry, wine, port and cognac were offered for the thirsty. The distinctions of rank were duly observed, with the ratings eating on the casing forward of the conning tower and officers remaining aft. It was a splendid example of how the French were able to come to terms with the horrors of total war.

The French minelaying submarine Rubis *in dry dock in 1941. Note the wells for the mines cut into her saddle tanks.* Rubis *served with great distinction with the Free French forces throughout the Second World War. (ECPA)*

Anglo-French relations during the Second World War were never easy. Churchill referred to de Gaulle as 'my Cross of Lorraine'. Yet in December 1944 when the Royal Navy released *Rubis* to French operational control, it was done with genuine regret. In his valedictory signal, Rear Adm George Creasy, Flag Officer Submarines, concluded, 'The absence of *Rubis* will be keenly felt not only by the 9th Submarine Flotilla but by the whole British submarine branch.'[4]

Modern submarines have a veritable arsenal of mines at their disposal. All mines are now designed to be laid from a standard 21in torpedo tube so that submarines can carry a mixed load of mines, torpedoes and SLCMs if necessary. Mines now come in two categories – the moored mine, similar to the German UC200 laid by the early UC-class, and the bottom mine. The moored mine consists of two canisters: the sinker, or weight that holds the mine on the seabed and then the warhead which, after separation from the sinker, floats up to a pre-set depth. The American Type 57 moored mine consists of 154kg of high explosive and can be laid in depths of up to 350m. Being only 3m long, the weapon occupies just half the space in a torpedo rack as that taken by a Mk.48 torpedo. A Los Angeles-class (SSN688) boat can carry forty-four such mines in the racks with a further four in each torpedo tube. However, the commanding officer would probably want to retain at least one torpedo salvo for defensive purposes if nothing else, so the maximum number of mines carried would be forty-four. What differentiates the American Type 57 mine from the German UC200 is that the introduction of micro-processing has turned a very dumb weapon, which had to be collided with before it would explode, into a very smart weapon that can differentiate between the acoustic signatures of different submarines. Detonation can be initiated by either acoustic or magnetic firing or, as a last resort, the traditional contact fuse.

A new development in mine technology is the tethered mine. Instead of reacting like a moored mine and exploding in position, the mine releases a homing weapon that takes the warhead to the target. The American CAPTOR mine releases a Mk.46 torpedo while the Russian Cluster Gulf fires a rocket-propelled torpedo when the target is in the mine's vertical attack zone. Sophisticated circuitry and micro-processing enable these mines to 'select' their target. Ground mines still have a role to play – the British Sea Urchin mine containing two 350kg warheads can be actuated by water pressure, magnetic or acoustic signatures.

Laying mines close inshore, or in shallow water where the enemy is active, is a demanding operation requiring precision navigation. Such an operation may require cautious use of the periscope with all the attendant risks of detection. The mines must be laid accurately, so that they do not constitute a danger to one's own forces, and laid cross-tide to make sweeping or clearance difficult. Several runs over the area are needed to complete a full field, even with the assistance of power loading. Loading the mines to keep to the laying schedule calls for very well-trained crews in the fore ends. Finally minelaying can cause temporary trimming upsets which will tax the watchkeepers to their limits. In shallow water the boat will not have much water under the keel but under no circumstances can the boat be allowed to broach. Finally there is the unwelcome but necessary requirement to break wireless silence to report completion of the operation.

Minelaying will doubtless remain a significant aspect of submarine operations. Mines are relatively cheap and can exercise influence out of all proportion to their size. The submarine remains the ideal carrier for these 'weapons that wait' because it is the most covert.

NOTES

1   Papers of Lt C. Allen RN, RNSM Archives.
2   MoD (Navy): British Mining Operations 1939–45, BR1736(56)(1) (London, 1973), p. 26.
3   PRO ADM199/3526 FS *Rubis* Report of Proceedings, 4 May 1941.
4   Admiralty: *Naval Staff History, submarines Vol. 1, Operations in Home, Northern and Atlantic Waters* (London, 1953), p. 222.

STRIKE FROM THE SEA

No one has done more to prevent conflict, no one has made a greater sacrifice for the cause of peace than America's missile submarine community.

Gen Colin L. Powell US Army on the return of the USS *Tennessee* from the 3,000th US Navy SSBN patrol, 25 April 1992.

Somewhere beneath the oceans of the world are lurking any number of American, British, Chinese, French and Russian ballistic missile submarines. They move extremely discreetly, avoiding all other shipping. Yet they operate on a war footing, remaining constantly alert for the signal from the Kremlin, the White House or the Elysée Palace ordering them to fire their missiles. The spectre of a submarine lurking beneath the sea, waiting to deliver a devastating attack using rockets or missiles, has been a part of submarine 'lore' for almost as long as the submarine has been in existence. In a sense the three British M-class submarines with their 12in guns were designed with such a role in mind – the surprise bombardment of enemy cities from the sea. One such operation proposed for *M.1* in 1918 was the bombardment of Constantinople. But it was the development of the rocket followed by that of the guided missile which turned this concept into a reality. Submarines armed with missiles now have the capability to deliver an attack many hundreds of miles from their launch position.

Submarines carry two sorts of missiles. First, there are Submarine-Launched Cruise Missiles (SLCMs), such as the American Sub-Harpoon and Tomahawk, French Sub-Exocet or Russian SS-N-12 Sandbox, which are used in a tactical way against other warships and land targets. They can be armed with a nuclear or conventional warhead and can be launched from either a conventional torpedo tube or specially fitted launchers. The former Soviet Union invested a massive amount into the development of such weapons, partially to compensate for American strike capability in the shape of carrier airpower. Second, there are Submarine-Launched Ballistic Missiles (SLBMs), such as the American Trident and French SNIAS M4. These are nuclear weapons of mass destruction which leave the earth's atmosphere for part of their flight and are targeted against capital cities, centres of population and concentrations of industrial facilities or raw materials. Their use is subject to the most stringent checks and controls and requires authorisation from the highest levels of government. A mutual exchange of such weapons would result in destruction on an unparalleled scale. (A third variety of submarine-launched missile, the anti-submarine missile, is dealt with on pages 175–6.

During the Second World War the Germans undertook some trials in the Baltic in 1943 firing short-range rockets from a submarine for tactical use against convoys but nothing much came of these plans. A more significant development came with a plan to launch V2 rockets against America. The weapon lacked sufficient range if fired from Europe but if towed to a suitable firing position, it could reach New York, Washington or any of the major population or industrial centres off America's eastern seaboard. The concept was to seal the V2 in a pressure-tight canister and tow it behind the submarine. Trials in the Baltic in 1943 in which a Type IX U-boat successfully towed a 90 ton cargo-carrying canister proved that the concept was viable. When the U-boat reached the firing position the canister would be ballasted so that it assumed a positively buoyant and vertical attitude and the weapon would be fired. Naturally the concept required calm weather and an absence of opposition during the pre-launch preparations. However, intriguing though this plan was from a German perspective, it remained on the drawing-board.

Viewed from the other side of the Atlantic, the spectre of a submarine-launched missile attack became horribly real in the spring of 1945. The FBI had captured German agents landed on the American east coast by *U1229* and *U1203*. Both these agents had indicated to the FBI that the Germans were planning to attack the US with submarine-launched missiles. This threat was taken seriously and on 10 December 1944 Mayor Fiorello LaGuardia warned New Yorkers about the possibility of a missile attack on their city. US concern about German intentions was heightened by the departure of a group of seven U-boats, *Gruppe Seewolf*, from Norway with orders to operate off the eastern seaboard. When added to intelligence about 'launching rails' observed on the casing of various U-boats in Norway and the fact that one of the boats heading west was commanded by KL Fritz Steinhoff, who while in command of *U511* had taken part in submerged rocket firing trials off Peenemunde, stories from captured spies suddenly acquired very real substance. In fact the thinking behind the despatch of *Gruppe Seewolf* was far more mundane. Dönitz believed that operations off the US coast and western Atlantic, where there was a lower concentration of ASW forces, might be more productive than in the English Channel and waters around the UK.

In response to this threat Operation Teardrop was launched: a massive activation of US coastal defences to combat any missile attack from the sea. However, the core of Teardrop was the unannounced deployment of two large carrier forces, each composed of one CVE and more than twenty DEs. Fortunately, HF/DF and decrypted German signals enabled Tenth Fleet to monitor the westward passage of *Gruppe Seewolf* so that Capt John R. Rushenberger's Northern Barrier Force, centred on the CVEs *Mission Bay* and *Croatan*, could position itself appropriately. Late at night on 15 April 1945 USS *Stanton* acquired a radar contact in thick fog. She closed the contact and when 1,000 yd off, illuminated it with a searchlight. From the bridge a wake and the dim shape of a U-boat could be seen. It was *U1235* which had surfaced since it was too rough to schnorkel. While *Croatan* manoeuvred out of the way, *Stanton*, now joined by *Frost*, carried out a Hedgehog attack. Six minutes later both ships were shaken by an explosion so violent that those in *Stanton* thought they had been torpedoed. The fog was now so thick that neither *Frost* nor *Stanton* could see the other, so their movements were coordinated by *Croatan*. *Stanton*'s third attack at 00.33 hrs produced an explosion so violent that even the *Croatan* (some 12 miles away) was violently shaken. The explosion was followed by the appearance of large and spreading slick of diesel.

Scarcely forty minutes after the sinking of *U1235*, *Frost* was in radar contact with another U-boat, *U880* (ObLtz.S Gerhard Schotzau), which was trying to escape on the surface, presumably having heard (and felt) the barrage that marked the end of *U1235*. The fog was now so thick that at 1,450yd, *Frost*'s starshell failed to illuminate the target. At 650yd the searchlight was switched on and the U-boat sighted. The seas were too heavy for *Frost* to alter course or bring all her guns to bear, but as the U-boat dived, *Frost* and *Stanton* followed her down with Hedgehog attacks. The two DEs remained in contact and at 04.04 hrs on 16 April there was a violent underwater explosion – as violent as that which had marked the end of *U1235* – followed by the spread of a diesel slick.

In the light of the fact that there were no survivors from *U880* or *U1235* and the fact that they were sunk so close together without leaving any identifiable wreckage, it is possible that the details of the sinking of these two craft are interchangeable. The third U-boat to be sunk in Operation Teardrop was *U518* (ObLtz.S Hans Offermann) which was detected on the sonar of the DE USS *Carter*, north-west of the Azores, which directed the *Neal A. Scott* in a creeping attack before attacking with Hedgehog at 23.09 hrs. One blast followed by several massive underwater explosions marked the end of *U518*. The very violent end of the first three boats sunk to date indicated that they might have been carrying more than just torpedoes. However, only the capture of survivors from one of these boats would confirm or deny whether a missile attack was imminent. The Americans' chance came with the sinking of *U546* (ObLtz.S Paul Heinz Just).

At daybreak on 24 April *U546* sighted an escort carrier (which was the *Bogue* of the Southern Barrier Force) being screened by DEs of Escort Division 4. Just was so intent on attacking the carrier that he attempted penetration of the destroyer screen at periscope depth, only to be detected by *Frederick C. Davis*. Just saw the DE turn towards him and fired a single T5 acoustic torpedo in a snap attack. The torpedo struck the *Davis* amidships and she sank very quickly with the loss of 126 of her 192 crew. The DEs now began a classic hunt using every attack method known to them. Just proved a wily opponent, going deep, using thermal layers and making constant alterations of course and speed while firing SBTs. After surviving repeated attacks for over ten hours, a Hedgehog attack by USS *Flaherty* caused massive damage forward and Just had no option but to surface. On the surface *U546* was engaged by virtually the entire Division before sinking. Despite having recently lost one of their number the DEs spared no effort to rescue the thirty-three exhausted survivors of *U546*, whose ordeal had been 'heard' by *U805*, creeping away as quietly as possible to the north.

Just (who had transferred to the *Kriegsmarine* from the *Luftwaffe* and who had flown in the Polish Campaign and the Battle of Britain) and his officers were questioned in the *Bogue*, where they had been transferred from the various DEs, but gave nothing away other than the information required by the Geneva Convention. On landing on 27 April at Argentia in Newfoundland, eight of the crew including Just were placed in solitary confinement and treated as military prisoners rather than POWs. They were then subjected to a rigorous exercise routine and repeated beatings in order to extract information on the rocket strike (which the unfortunate U-boat men knew nothing about). This treatment continued despite the fact that the USS *Varian* had recovered a diary from a *U546* survivor which showed that the boat was armed with nothing more than the usual mix of contact, acoustic and homing torpedoes. *Varian*'s CO, Lt Cdr Leonard A. Myhre USN, was invited to witness one of Just's 'exercise' sessions and protested bitterly at what he saw. Possibly on account of his protest, the men were flown to Fort Hunt near Mount Vernon where the 'exercise' and beatings resumed. This treatment only ceased just after VE Day when Just agreed to write a full account of *U546*'s operational history. Nevertheless, the Americans remained concerned. When KL Fritz Steinhoff surrendered his boat to the Americans at the end of the war, he, like Just, was taken into FBI custody rather than treated as a POW. Steinhoff was treated so badly that eventually he took his own life in Boston's Charles Street prison.

After the war the Americans 'liberated' a significant amount of German rocket technology and personnel and it was not long before the US Navy was experimenting with launching such weapons from a submarine. A missile-armed submarine gave the ability to deliver a strike against the enemy that was all but undetectable until the very last moments. Unlike airfields or land-based missile sites, submarines were not as vulnerable to being destroyed in a first strike by the opposition. The missile first used was the Loon (nothing more than a copy of the German V1) which was launched for the first time from the USS *Cusk* (SS348) on 12 February 1947. The weapon was mounted on a ramp on the after casing, although shortly after conversion both *Cusk* and the *Carbonero* (SS337), which was similarly fitted, were given pressure-tight hangars on the casing in which to keep the missile. Loon required mid-course guidance so a number of other submarines were fitted with guidance equipment to pick up and direct the missile while in flight.

The Loon programme was terminated in 1953 and replaced with Regulus, a much larger missile which resembled a pilotless aeroplane. The submarines *Carbonero*, *Tunny* (SS282) and *Barbero* (SS317) were converted to carry the missile. The conversion involved the removal of one main engine and the auxiliary engine, the installation of a missile hangar, a retractable launching platform and considerable internal re-arrangement in order to accommodate missile control and servicing equipment. Both Loon and Regulus had one fundamental disadvantage. To fire the weapon the submarine had to surface and then the missile had to be removed from the hangar and placed on the launching ramp prior to firing. Bad weather could complicate this task enormously. Moreover there was always the fear that the submarine could be caught on the surface during the lengthy firing drill. Regulus was eventually supplanted by the Polaris programme and US interest in air-breathing missiles lapsed until the development of the Tomahawk.

The Tomahawk is a versatile weapon. It is carried by American ships, aircraft and land forces as well as submarines. It can be carried in the standard 533mm torpedo tubes of the *Sturgeon* and *Los Angeles* (SSN-688) SSNs although all boats of the latter class from SSN-719 onwards will be fitted with twelve dedicated Tomahawk launchers. Tomahawks will also soon be carried by British SSNs of the Trafalgar and subsequent classes. Tomahawks can be fitted with either a nuclear or conventional warhead according to the mission parameters.

Tomahawks can be programmed for either the anti-ship or land strike roles. For the former the missile carries a 454kg conventional warhead (following the decision taken in 1994 to withdraw all tactical nuclear weapons from US ships and submarines) and uses an inertial system to find the target. Pre-set values (the firing ship's position and the target's likely future position) are fed into the missile's computer and it then uses its own computer to make its way to the target. The information can come from a variety of sources, including other ships and satellites, as well as the submarine's own sensors. Once in flight the missile can also home in on the target using its own radar or the target's radar emissions.

For the land attack role the missile uses DSMAC (Digital Scene Matching Area Correlator) to find precisely defined targets. This consists of a miniature TV camera fitted in the missile's nose which records the passage of landmarks and topographical features while the missile is in flight. These images are then compared with similar images stored in its memory and adjusts the inertial system as required to keep the missile on course.

In the 1991 Gulf War two US SSNs, USS *Louisville* and USS *Pittsburgh*, successfully bombarded targets inside Iraq using the Tomahawk cruise missile from positions in the Red Sea and Eastern Mediterranean respectively. In the future Tomahawks will increasingly be used as a substitute for air attack. First, the missile has an all-weather capability which, when combined with impressive accuracy, means it can deliver a strike on a given target with minimal collateral damage. Second, such an attack does not run the risk of losing lives and highly expensive aircraft – and the subsequent humiliation of having the aircrew displayed (dead or alive) on the opposition's television. As the US government seeks to take a tougher line with fundamentalist Islamic terrorists, more use can be expected of this weapon.

Other western weapons are the Sub-Harpoon and the Sub-Exocet, both variants of highly successful weapons already fitted to ships and aircraft. Sub-Harpoon is the most ubiquitous of the two, being carried by US, British, Australian, Greek, Israeli, Japanese, Dutch, Turkish and Pakistani submarines. Italy and Taiwan have also expressed an interest in the weapon. Sub-Exocet is only carried by French SSNs and Agosta-class SSKs in French and Spanish service.

The launch of a Tomahawk cruise missile fired by USS Louisville *during the 1991 Gulf War. The photograph was taken through the periscope while the submarine was dived. (US Navy)*

Early Soviet efforts in this direction anticipated these developments. The Russians had liberated almost as much ex-German rocket technology as the Americans and were only slightly slower in putting it to use. The SS-N-3 Shaddock missile-fitted in the Whiskey-class SSGs. Shaddock was stowed in tubes which could be hydraulically raised and lowered from inside the boat thus removing the need for anyone to be on the casing during the pre-launch drills. Once fired the missile's wings deployed automatically, removing another task which in the Regulus had to be done by the crew. However, the missile tubes greatly compromised streamlining, a problem which continued to dog subsequent Juliet SSG and Echo II SSGN conversions.

Thereafter the Soviet Union produced a series of SLCMs. Shaddock was successively improved into the SS-N-12 Sandbox and SS-N-19 Shipwreck which were long-range anti-ship missiles. A significant step forward came with the introduction of the SS-N-7 Starbright which was the world's first sub-surface launched, autonomous, tactical anti-ship missile, fitted to Charlie-class SSGNs in 1967. The Russians have also been quick to follow the USA with the development of the SN-N-21 Sampson and SS-NX-24 long-range terrain-following cruise missiles. Russian SSGNs such as the Oscar, the US Navy/Nato designation for the very capable Soviet Project 949 Antey cruise missile submarine, carry a formidable number of these weapons: twenty-four SS-N-19 cruise missiles, two 533mm and four 650mm tubes with twenty-four tube-launched torpedoes. Russian thinking behind such weapons is clear. They free a submarine commander from the traditional constraint of having to approach to within a relatively close distance of the target before firing. Furthermore, the launch of these weapons in salvoes may well overwhelm a carrier group's or convoy's anti-missile defences.

The early Russian missiles such as Shaddock, Sandbox and Shipwreck used command-guidance where the submarine established contact with an aircraft (hence the Russian requirement for a very large fleet of long-range maritime patrol aircraft such as the Tupolev Bear) before firing and used the aircraft's radar to guide the missile to the target. Subsequently the Russians adopted inertial systems and TV guidance.

The *modus operandi* of SLCMs, of whatever nationality, is basically the same. Missiles can, after an initial bunt to an altitude of 300m, follow a sea-skimming flight profile using highly accurate radio altimeters with the intention of getting beneath the enemy's radar until the missile gets so close that avoiding action is impossible. Alternatively the missile can climb to a very high altitude, well above the target's defensive weapon engagement envelope, at speeds of Mach 2 plus before plunging vertically down on to the target.

SLBMs are a different class of missile altogether. As the Cold War intensified, America and Russia commissioned extensive nuclear forces with the capacity to destroy the other side several times over. However, on both sides there was the lurking suspicion that these forces might be destroyed in a pre-emptive first strike by the other, so almost as much effort went into protecting the missiles from attack as went into their design and construction. The sea offered the best chance of concealment and the recent development of the nuclear-powered submarine offered a 'carrier' with almost unlimited submerged endurance.

Table 2: Soviet Submarine-Launched Anti-Ship Missiles

Missile Code Name	Range (nm) Speed M	Guidance Homing	Warheads HE/Nuc	S/Ms (Nos) Missiles Carried
SS-N-3C Shaddock	460 1.1	Command Radar	1,000kg HE 350KT Nuc	Juliett (4) 4 each Echo 2 (2) 8 each
SS-N-12 Sandbox	550 1.7+	Command Radar	1,000kg HE 350KT Nuc	Juliett (8) 4 each Echo 2(12) 8 each
SS-N-19 Shipwreck	620 1.6	Command Anti-Radar	HE or Nuc	Oscar (9) 24 each
SS-N-17 Starbright	64 0.9	Aimed Radar	500kg 200KT Nuc	Charlie 1 (8) 8 each
SS-N-9 Siren	110 0.9	Aimed Radar & IR	500kg HE 200KT Nuc	Charlie 2 (6) 10 each
SS-N-21 Sampson	2000 0.7	Inertial TV Match	200KT Nuc (CEP 150m)	Yankee (3) Akula (9) Sierra (3) Victor III (26)
SS-NX-24	4000	Inertial	1 MT Nuc	Yankee Conversion 12[1]

In 1957 Vice Adm William F. Raborn was given the task of developing a totally new kind of missile and sending it to sea in a new type of submarine. Space precludes an account of the development of what was to become the Polaris system but on 20 July 1960 Raborn stood in the control-room of the USS *George Washington*, which was armed with twenty-four Polaris A1 missiles. He then watched as she hurled two missiles with perfect accuracy down the Atlantic Missile Range. His achievement was considerable, equal to that of Rickover, yet accomplished with hardly any of the storms or verbal histrionics which accompanied the US Navy's nuclear power programme. As one contractor put it, 'He doesn't yell. He doesn't snap at you. He just smiles and asks you one damned question after another about things you thought he knew nothing about.'[2] Polaris A1, a solid-fuelled rocket with a range of 1,200nm, first flew in 1960 and was fitted in the five George Washington-class SSBNs. A1 was replaced by the A2 (1,555nm range) which was fitted in the Ethan Allen- and Lafayette-class SSBNs. Intelligence reports about Soviet anti-ballistic missile defences led to the introduction of the A3 variant in 1963. The range had gone up to 2,500nm but in order to confuse Soviet defence

USS George Washington, *the first SSBN. The arrangement of the missiles in a section aft of the fin is now standard among nearly all SSBNs regardless of nationality.*

systems, the missile carried three warheads. It was the A3 version that was supplied to Britain in 1962 under the Nassau agreement for fitting in Resolution-class SSBNs.

Polaris was in turn replaced by Poseidon which was fitted in thirty-one American boats. It has a range of 3,230nm with ten 50KT warheads. However, looming behind Poseidon was the Trident system. Trident is a three-stage solid fuel rocket approximately 13m long and weighing 60 tonnes. Each missile can deliver twelve warheads from a MIRV, together with a number of decoys in order to confuse ABM defences. To further confuse ABM defences the actual warheads are disguised to resemble the decoys. The defence problem can be made worse by using Penetration Aid Packages (PAPs) which create a large number of false targets. These PAPs burn up on re-entry but the defences, which should fire two ABMs at every target in order to cover a malfunction, must start the engagement while the PAPs are still exo-atmospheric. This overloading of the ABM system is known as 'winning the exchange ratio battle'. Defences will be further stretched by the fitting to Tridents of MIRVs. Trident is fitted in the American Ohio- and British Vanguard-class SSBNs.

The United States has ten Ohio-class Trident D5 SSBNs in commission and four (possibly eight) of the Trident C5 boats will be retrofitted to take the D5 missile in due course. Each D5 SSBN carries twenty-four missile tubes – each missile carries twelve thermonuclear MIRVs.

Soviet developments in this field took longer to reach fruition. Early Soviet missile boats were diesel-powered, used a variant of the Army's Scud missile and had to surface to launch the missile. The first true Soviet SLBM was the SS-N-4 Sark missile which was deployed in the Hotel-class SSBNs (or PLARBs in Soviet parlance). Unlike western boats, where the missiles were carried in the hull proper, the Hotels carried their missiles in an elongated fin. The first Soviet PLARB to equal western designs was the Yankee which carried sixteen SS-N-6, later replaced by the SS-N-8. The SS-N-8 has a range of 4,250nm and can be fitted with either a single 2.5MT warhead or MIRV capsules.

Unusually SS-N-8 is powered by liquid fuel. Undoubtedly it was an explosion following leakage of liquid fuel that was responsible for the loss of a Yankee-class SSBN, *K-219*, in the Atlantic 480 miles east of Bermuda on 3 October 1986. The seal on one of the missile tube hatch covers failed, allowing sea water to leak into the tube. Some reports, unconfirmed, indicate that the leak was triggered following a near-collision with a US Navy SSN which was shadowing *K-219* at the time. The sea water then reacted with fuel leaking from the missile, causing an explosion. Three crew members were killed in the explosion. The submarine then began to experience problems with her reactor which eventually had to be shut down by hand – one member of the crew sacrificing his life to do so. Eventually, after all efforts to save the submarine failed, her crew were taken off and the boat scuttled in 18,000ft of water about 600 miles north-east of Bermuda.

SS-N-8 was fitted in Yankee- and Delta-class PLARBs. In response to the deployment of Pershing II and cruise missiles in Europe in the late 1970s, the Soviets deployed the SS-N-18 missile in the Delta III variant PLARB. SS-N-18 has three variants: MOD-1 which is a 200KT MIRV vehicle with a range of 5,988nm; MOD-2 carrying a single 450KT warhead to a range of 4,970nm; and the MOD-3 carrying seven MIRV warheads of an unknown yield. Finally the SS-N-20 missile is fitted in the Typhoon-class SSBNs.

Typhoon is the western designation for the Soviet Project No. 941 which has a displacement of 18,000/25,000 tons and a length of 171m, making them the largest SSBNs in the world. The components of the submarine are so large that when work first began on their assembly at Severodvinsk in 1977 US intelligence analysts thought the Russians were starting work on an aircraft carrier. Armed with twenty SS-N-20 missiles, these boats are designed to sit out the initial nuclear exchange under the ice and to be available for a reprisal attack on a recovering enemy. Hence the extraordinary emphasis placed on habitability for the crew of 50 officers and 100 men – there is even a sauna. The exterior pressure hull conceals five cylinders: two main units running the length of the boat, one forward for the six torpedo tubes and tactical missiles; one for command located under the fin; and one aft for steering. Each of the two main hulls contains a 190mW reactor, a 45,000shp turbine and an 800kW diesel

The British SSBN Repulse *moves away from the Clyde submarine base in 1973. (MoD)*

generator. Unusually in SSBN design the strategic missiles are placed forward of the fin. Six of the class are in service.

Russia currently has thirty-six PLARBs in various states of availability – although only 50 per cent are reported as fully seaworthy and this number is expected to reduce to twenty-six as the Russian Navy continues to contract. PLARBS are equipped with various missiles which are not as accurate as Trident but which will be replaced toward the end of the century by a new missile of improved accuracy which will be retrofitted to existing boats. The ability of the shipbuilding industry as currently constituted to support this programme is in doubt. However, the decision to concentrate all submarine construction and refit work in Severodvinsk under tight management control from Moscow will improve matters.

Britain signed up to the SSBN programme extremely quickly, following agreements reached at Nassau by Prime Minister Harold Macmillan and President John F. Kennedy. Personal links between the two men were critical in allowing Britain to purchase what was the most sophisticated piece of military hardware in the American arsenal. Britain bought the Polaris system from the USA to fit in the five (later scaled down to four)

Resolution-class SSBNs. The British Polaris programme was managed by Vice Adm Sir Hugh Mackenzie, a distinguished wartime submariner, who found himself unexpectedly and hurriedly relieved of his appointment as Flag Officer Submarines in order to manage the new project. After being told of his new appointment, 'I found myself with my dog in an empty office in the Admiralty with just a chair and a wastepaper basket, not even a telephone . . . this was the very low key start of the Polaris programme.'[3]

Despite such an inauspicious start the British Polaris programme was a resounding success. The four submarines were built on time and within budget – a unique achievement for British industry. Britain has since retired her fleet of four Polaris SSBNs and now has three (plus one under construction) Vanguard-class Trident D5 boats in commission. The British boats carry twelve missiles and the British warheads are a different design to the US variant. However, the British government has indicated that it will not use the full capability of the system and that each British Trident SSBN will carry a maximum of 128 warheads. This would seem sufficient for British needs: Trident has a range of 4,000 miles and there is no place on earth more than 1,400 miles from the sea.

France became the fourth power to commission SSBNs (known in France as SNLE) for reasons of national prestige. If France were to maintain her place at the 'top table', she needed the same capacity to inflict mass destruction as the other leading powers. The SNLE programme was carried through at great cost to the rest of the French Navy. The first French missile, the M1, had a range of 1,350nm and was later replaced by the upgraded M2, M20 and M4 which is about to enter service. M4 is a MIRV system of six individual units each with a 150KT warhead and a range of 2,485nm. France resolutely demonstrated its commitment to nuclear weapons in 1996 by carrying out live nuclear tests in the Pacific against a background of strident international protest. The new Le Triomphant-class SNLEs are taking longer to build than their British and American counterparts and the boats of the older L'Inflixible-class will remain in service to be paid off one by one as their new replacements enter service. The cost to both Britain and France of the SSBN programme has been enormous and cannot be counted in mere financial terms. Both the Royal Navy and the Marine Nationale have had to fund their deterrent programmes out of purely naval budgets, with a knock-on effect on other naval programmes. The 'conventional' naval forces of both countries have suffered as a result.

Lastly, we come to China. The Chinese government has demonstrated an unrelenting determination to commission an SSBN, but so far technical problems have dogged the sole Xia-class SSBN to commission, while a second was reported as lost in an accident. However, Russian assistance, paid for in hard currency, will see the Chinese join the 'SSBN club' as a fully paid up member within the next couple of years.

Three considerations dominate SSBN operations: secure communications; quick reaction to the receipt of a firing message and remaining undetected. All three are self-evident. An SSBN on patrol must be able to receive signals from its government. SSBNs usually communicate on VLF or ELF and considerable effort and expense has gone into making sure that there are back-up transmitters ashore in order to maintain the link. It

HMS Victorious, *second of Britain's four Trident SSBNs, is rolled out of the building shed at Barrow in 1996. The men in the foreground give an idea of the impressive size of these submarines. Note the four apertures, two either side, for the 21in torpedo tubes. (VSEL)*

also follows that an SSBN's own communications equipment is kept fully serviceable at all times, particularly during periods of political tension. SSBNs also need to be able to monitor domestic radio traffic in case of a surprise attack on the homeland which might have destroyed the government as well as their normal transmitters. It is reported that British SSBNs listen to BBC Radio 4's *Today* news programme as an indication that 'All's Well'. Presumably American, French, Russian and Chinese SSBNs have similar routines. The consequences of a breakdown in communications were grimly, if somewhat theatrically, depicted in the feature film *Crimson Tide* (1996), in which communications failure led to the XO of an American SSBN refusing to support his CO in the decision to order a release of strategic nuclear weapons. On a more serious note, the effectiveness of a deterrent is reduced if the enemy knows that communications are unreliable or indifferent.

Once a message is received it must be acted on immediately. The dreadful consequences of a nuclear exchange must be put aside. Any government, be it British, French, Russian or American, must have the confidence in its SSBN crews that once the political decision has been taken to resort to nuclear attack, then the order will be followed through.

Finally, SSBNs must remain undetected to ensure their effectiveness. If the opposition can tail an SSBN, then it can be taken out in a pre-emptive strike and the shore command might not receive word of the boat's loss and believe her to be still active. This is of particular concern to Britain and France who maintain only one SSBN on patrol at any one time. Discovery of that SSBN would effectively permit blackmail of the government concerned. Thus considerable effort goes into 'delousing' SSBN transit routes to patrol areas.

SSBN operations are unlike any other kind of submarine activity. Patrol routines and refit programmes mean that a member of an SSBN crew can forecast his programme for months ahead, unlike any other member of a uniformed naval service who is liable to be sent to sea at a moment's notice. While on patrol, life is a puzzling combination of an advanced state of readiness and daily routine. At all times the submarine must be prepared to fire her missiles on receipt of the appropriate signal. Part of this readiness includes the ability to move away from ships/submarines of any nationality that enter the SSBN's area. Frequent drills are held with 'dummy' firing messages to ensure that missile launch can be achieved within the given period. Such drills are recorded and the results exhaustively analysed afterwards.

On receipt of a firing message the boat is brought to the optimum firing depth and speed for launch of the missiles. Meanwhile, the commanding officer, his executive officer and the weapons officer go through an exhaustive series of checks to establish the authenticity of message. Unlike the firing of a torpedo where the commanding officer has sole discretion, the firing sequence for an SLBM requires the public concurrence of another senior officer on board. On western boats that would be either the executive officer or the weapons officer. In former Soviet boats it would undoubtedly have been the *Zampolit*. Every order given by the commanding officer is relayed throughout the boat on the submarine's main broadcast system and immediately repeated by the authenticating officer. Finally, to actuate the firing circuits in the system requires that two firing keys be simultaneously turned in locks in different locations in the submarine by different officers. The system is designed so that no single officer can initiate a launch on his own authority.

As land and air-based nuclear missiles decline in numbers, more intercontinental ballistic missiles are being carried in submarines. It is the cheapest and most secure form of maintaining a deterrent. Great Britain, France, the United States, China and Russia have reaffirmed their commitment to seaborne nuclear deterrence even though the Cold War ended in 1992. Whatever their reasons, it is a unanimous vote of confidence in the continuing strategic significance and invulnerability of these weapons and of the nuclear-powered submarine as the preferred platform.

NOTES

1   Specifications from Hervey, Rear Adm, J., *Submarines* (London, Brasseys, 1994), p. 132.
2   Stafford, Edward P., *The Far and the Deep* (London, Arthur Barker, 1968), p. 348.
3   Vice Adm Sir Hugh Mackenzie to author, 7 February 1989.

ESCAPE AND RESCUE

The issue of escape and rescue is inseparable from that of submarines. Ever since man first went underwater, he has been faced with the problem of how to extricate himself from a sunken submarine. There are two main causes of submarine accidents: human error and failure of *matériel*. The former encompasses the full gamut of human frailty, from criminal negligence to sheer bad luck. For example it was carelessness and failure to observe the basic principles of submarine management that caused HMS *Artemis* to sink alongside the jetty at HMS *Dolphin* on 1 July 1971. On the other hand, the American submarine *S.51* was run down and sunk by the Italian liner *City of Rome* in an accident in which the submarine was not at fault. Failure of *matériel* includes breakdowns, the failure of hull valves, metal fatigue and a host of other defects that can plague even the best regulated submarine. In May 1939 USS *Squalus* sank in the Gulf of Maine when an engine induction valve failed to close as the submarine dived. The two causes are not exclusive: occasionally they come together in a cruel combination.

The history of submarine escape has swung like a pendulum swinging between two different philosophies: first, giving the crew the means whereby they can leave a stricken submarine without outside assistance and, second, building up a salvage organisation that can raise the submarine and her crew intact. For the first part of this century rescue by salvage was the preferred method. However, developments in escape techniques – and some lessons learned the hard way – meant that by the end of the Second World War submariners of all nations were taught to escape rather than wait for rescue. Generally speaking, the advocates of rescue have had the upper hand over the salvage lobby. However, providing a submarine with rescue equipment implies a fundamental contradiction in the boat's design. Submarines are military machines designed to have the greatest efficiency in war. Lavish provision of escape equipment must inevitably reduce fighting potential. Something less than the ideal must be accepted and the first priority given to sound design and construction along with high levels of compliance with design and training requirements. Nevertheless, all navies have devoted a good deal of investment in this area, although it must be said that, particularly in the case of the Royal Navy, it sometimes took a submarine disaster in order to persuade a reluctant government into authorising the necessary expenditure.

All escapes from a submarine involve one fundamental principle. The submarine, or a compartment in the submarine isolated from the rest of the boat, has to be flooded up so that the pressure inside the hull equals the sea pressure outside. It is then possible to remove the clips from a hatch and leave the submarine. This was the method used in the first genuine submarine escape on 1 February 1851 when three men escaped from the German *Brandtaucher* which sank off Kiel. Wilhelm Bauer, the boat's designer, was on board and after failing to get the boat to rise by blowing the ballast tanks he convinced his two companions that the only way to escape was to flood the craft with seawater until the pressure equalled and then to open the hatch and swim up to the surface. This is what happened. 'We came to the surface like bubbles in a glass of champagne,'[1] said Bauer afterwards. The *Brandtaucher* was salvaged and now resides at the Mariners' Museum in Newport News and Bauer's place in the history of submarine escape was assured.

The one disadvantage of Bauer's method is that while the flooding of a compartment is taking place prior to an escape, the human body is placed under pressure. The greater the depth at which the submarine lies, the greater the pressure will be. The length of time to which men are exposed to raised pressures is critical in attempting escape from any but shallow depths. Men exposed to raised pressures of gas will absorb an increased quantity of gas into the blood and tissues which causes decompression sickness when pressure on the body returns to normal. Flooding the compartment also increases the partial pressure of all gases present including CO_2 and other gases such as CO and chlorine which are particularly dangerous. The effect of a given level of CO_2 is relative to the depth at which the submarine is lying. The time taken to reach the maximum safe level will depend on the volume of the space, the number of men present and their level of activity.

Because of these constraints, rescue by salvage was the preferred method of escape for much of the early part of the century. Predictably it was the Germans who had given most thought to the matter. At this period submarines were considered to be coastal creatures who would not venture far from port and rescue facilities. The first German U-boats were equipped with telephone buoys which people on the surface could use to communicate with a submarine below. The Germans also completed two submarine rescue ships, *Vulcan* launched in 1907 and *Cyclop* launched in 1916. Both ships were, in effect, catamarans equipped with lifting apparatus to raise a submarine between the two hulls. Similar ships were built by Russia; the *Volkhov*, launched in 1913, is still extant! In 1913 *Vulcan* successfully raised the *U3*. However, it took eighteen hours to get *Vulcan* to the scene and in position over *U3*, and to bring the submarine to the surface. By the time the rescuers forced the hatch open only three of *U3*'s crew were still alive and all three died not long afterwards. This episode illustrated a fundamental problem – by the time the salvage operation was completed those inside the submarine would be dead of CO_2 or chlorine gas poisoning.

During the First World War and in the interwar period there were a number of successful escapes from submarines where rescue was by salvage. In January 1917 forty-six men were rescued from the British submarine *K.13* which sank in the Clyde while on trials. A few weeks previously five men had been rescued from the Danish submarine *Dykkeren*. On both occasions salvage ships, equipment and expertise were close at hand. Two submarines were lost in the 1920s, *S51* in 1925 and *S4* in 1927. Salvage efforts on both boats failed to save the crews and there was considerable public comment – nearly all of it bitterly critical of the US Navy. The Americans regrouped by developing the McCann Rescue Bell, a diving bell with two compartments, which could be lowered down on to the hatch of a sunken submarine. By using the lower compartment as an airlock the submariners could climb up to the upper compartment without even getting wet and while remaining in atmospheric pressure. The Bell could then be hauled to the surface and safety. Undoubtedly, the most significant success of the Bell was in May 1939 when thirty-three men were rescued from USS *Squalus*, which had sunk in the Gulf of Maine.[2] This was an outstanding rescue but depended on the proximity of the salvage ship *Falcon*, which carried the Bell.

There were other problems associated with the use of the Rescue Bell; calm seas and an absence of strong tides and currents were essential. The Bell could also only be used at depths in which divers could work, since a diver had to secure the down wire, on which the Bell travelled, to the sunken submarine. In 1953 the Turkish submarine *Dumlupinar* sank in 276ft of water in the Dardanelles after being rammed by a Swedish merchant ship. The Turks had purchased the McCann Bell, largely on the strength of the *Squalus* rescue, and their salvage ship, with the Bell in position, was swiftly on the scene. However, divers found it impossible to secure the down wire owing to the strong currents. The Royal Navy carried out trials with the McCann Bell in the postwar period but ultimately rejected it largely because it would be impossible, or at best prohibitively expensive, to maintain a world-wide network of Bells available for immediate use in a given location.

The alternative to salvage is for submariners to conduct their own escape using equipment from within the submarine. This poses two problems: first, how to leave the submarine, which involves the provision of escape hatches or escape chambers; and second, the design and provision of a suitable breathing apparatus to allow the wearer to breathe during the flooding up process and the ascent, and also to keep him afloat on the surface afterwards.

How to leave the submarine? All submarines are fitted with hatches and early escapes usually took place through one of these hatches. A torpedo tube presents another way out. In 1909 Ensign Kenneth Whiting USN left the submarine USS *Porpoise* via her torpedo tube. As the tube flooded and the bow cap opened upwards, Whiting grasped the strongback and allowed himself to be pulled out of the tube, from where he swam to the surface. Other US officers performed the same experiment, while Midshipman Deem not only left a submarine via the torpedo tube but re-entered a submarine at a depth of 40ft by the same means. In 1949 the Dutch naval officer Lt Cdr Jan Lenderink RNethN made a highly publicised exit from a Dutch submarine via a 21in torpedo tube. It took over three minutes to flood the tube before the bow cap could be opened and Lenderink 'freed'.

Torpedo tube escapes have many disadvantages. It is self-evident that the last man must sacrifice his own life since he has no means of closing the rear door of the tube behind him. Damage to the bows or the bow cap operating machinery may make it impossible to open the bow cap. The submarine's bows may be buried several feet deep in mud at the bottom. And not every man can squeeze into a 21in diameter tube! Lastly, men like Whiting and Lenderink made their escapes coldly and deliberately under controlled conditions. In a real disaster there might well be men who could not face the ordeal (or who might be wounded) of climbing into a black, cold and wet tube, having the door closed behind them and then waiting while the tube was flooded up. There has only been one instance of a torpedo tube used as an escape route in a real disaster. On 16 April 1917 the British submarine *C.16* (Lt Harold Boase RN) sank in Harwich harbour following a collision. The first lieutenant, Samuel Anderson, crammed himself into an 18in torpedo tube in an attempt to get out of the submarine and summon help, but he perished in the attempt. Either Anderson was too big to get out of the tube or the submarine was lying at a steep bow up angle.

During the First World War there were a number of escapes from submarines. On 14 July 1916 four men escaped from *U51* which had been torpedoed by the British *H.5* off the Ems Estuary. Most notable of all was the escape of Stoker PO W. Brown who was trapped alone in the engine-room of the submarine *E.41* which sank after colliding (on the surface) with her sister ship *E.4* on 15 August 1916. Brown had remained in the engine-room to check that everybody had got out and when he reached the control-room he found it deserted and the hatch shut. He then returned to the engine-room and set about making his escape through the engine-room hatch. This was no easy task since the hatch opened inwards and was kept in place by a heavy steel strongback, which was normally secured by two or three men before the boat went to sea. Nevertheless, in order to increase the pressure, Brown opened every valve he could find in the engine-room, which was now lit solely by some failing emergency lamps,. The increase in pressure forced the hatch up on its seating and thus Brown could take off the strongback and securing clips.

However, his problems were just beginning for the sea pressure outside was forcing the hatch open and his 'airlock' was disappearing. He had to replace some of the securing clips (which he had just removed) by diving into the oily bilges and groping for them. Eventually he found two and resecured the hatch. Eventually with the water up to his shoulders he removed the clips and tried to move the hatch. But, the hatch merely gave a little and then snapped shut. He waited, then tried a second time, only for the hatch to snap shut on his hand. Somehow he mustered enough strength to move the hatch sufficiently to release his hand. He now resolved to wait until there was no airlock left – the last moment when he could hope to move the hatch. Accordingly he waited until the water level in the engine-room was so high that he could barely keep his nose and mouth above it, before making one last effort. This time he was successful and he shot to the surface where he was picked up by the destroyer HMS *Firedrake*. Brown's achievement was courage of the highest order.

These escapes from *U51* and *E.41* were desperate affairs, carried through solely by the determination of those involved. However, these incidents did provide the stimulus for some developmental work between the wars. The first (like so many, of uncertain date) was the development of an inverted coaming or trunk projecting downwards into the boat from the escape hatch. The problem with the escapes undertaken by those in *U51*, by Stoker PO Brown in *E.41* and by the men trapped in HMS *Poseidon*[3] in June 1931 was that having flooded up and equalised the pressure, when they opened the hatch all the air disappeared at once in one big bubble. In the case of *Poseidon*, having waited for *2½ hours* to flood up the compartment, only two men got out of the submarine when the hatch was first released. The hatch then slammed shut and it took a *further hour* to equalise the pressure to allow the remainder to escape. During this period the six remaining men were in acute discomfort, clinging to a wire hawser strung directly beneath the hatch in a compartment filled with cold, oily water and without lights.

Designers in America (Momsen) and Italy (Belloni) developed the same solution at roughly the same time. This consisted of a solid skirt which was let down from the escape hatch in order to form a column of water in the submarine. This trunk ensured that a pocket of air was left in the boat when the hatch was opened, for without it all the gas would escape immediately. Momsen's original design called for a solid steel trunk but this was obviously impractical. In the Italian design, and in British practice, where it was known as a Twill Trunk, it comprised a collapsible canvas structure suspended from the hatch coaming and secured to eye plates on the deck. To conduct an escape using this structure, it was necessary to lower it to the right length (as indicated by a graduated scale on the trunk corresponding to the depth in which the submarine was lying and then flood up the compartment. When pressure was equalised – as shown on a differential depth gauge, a seaman would climb up inside the trunk and release the air in the trunk through a vent in the upper hatch. The trunk then flooded and the hatch could be opened. All that was necessary to leave the submarine was for the men to duck under the hatch one by one and head for the surface. The Twill Trunk was a partial solution but it still exposed the men to great pressure while the compartment was being flooded. Accordingly, Britain and America began to develop dedicated escape chambers holding two or four men at a time. These chambers were smaller so compression time was quicker and the men were exposed to less pressure.

At the same time progress was made in the development of breathing apparatus for use in submarine escapes. Such equipment was developed in the early part of the century and was based on safety equipment then in use in the mining industry. All such equipment – the German Dräger breathing apparatus introduced in about 1912, the American Momsen Lung introduced in 1928 and the British Davis Submarine Escape Apparatus introduced in 1929 – works on the same principle. The set consists of an oxygen bag, worn around the neck, which is filled from a small cylinder. The wearer breathes oxygen from a mouthpiece and the used air is then recycled through a cartridge containing a CO_2 scrubbing agent. The oxygen which is absorbed by the body can then be replenished

British seamen examine a German Dräger breathing apparatus recovered from U357, which was rammed by HMS Hesperus *on 26 December 1942. (Author)*

from the cylinder. The bag acts as a buoyancy aid during the ascent and as a life-jacket on the surface. The most significant difference between the sets is that the DSEA and Dräger sets are self-contained while the Momsen Lung can be recharged from a 'ring main' of oxygen bottles in the escape tower. All types of breathing apparatus had to be small enough so that sufficient sets for every man on board could be stowed in the submarine without taking up too much room. Bernard Dräger, designer of the set that bears his name, wrote: 'It seems essential that a life-saving apparatus must have exceptional qualities if it is not to be superfluous ballast in the restrictive space of a submarine.'[4]

Unbelievably, there were some people who disagreed with the whole idea of submarine escape. In November 1934 the British Director of Plans, Capt E.L. King RN (a non-submariner), commented: 'DSEA should not be carried in wartime. The availability of such gear is calculated to precipitate a panic during a depth-charge attack. If there is no escape apparatus, each man will know that his only chance in danger is to do his work and obey orders.'[5]

A more imaginative way out of a sunken submarine came from Italy where the Gerolami-Arata-Olivati system was unique, if nothing else. A cylindrical chamber, large enough to hold one man, was stowed in a

vertical tube which ran from top to bottom in the submarine's pressure hull. At the lower end the hatch into the chamber was attached to a wire running round a pulley to a winch. At the upper end the trunk was sealed by a hatch. To escape a man would enter the chamber through a watertight door in the cylinder which would then be closed behind him. The trunk would then be flooded up and the upper hatch opened. The brake would be taken off the winch, after which positive buoyancy in the cylinder would take it to the surface. There the man stepped out of the chamber into the sea or a rescue boat. The cylinder was then hauled back to the submarine and the operation repeated. This most ingenious of devices was known as the *Ascensio Submarino*, and a variant was fitted in some Spanish submarines. However, it was never generally used because it took up too much space, added weight and inevitably formed a weak spot in the pressure hull. However the idea of an integral yet detachable escape chamber has not entirely died, see page 129.

The British T-class built from 1937 onwards seemed to embody the most comprehensive escape arrangements possible. Each submarine carried DSEA sets for every man on board plus a small reserve. Each submarine was also fitted with two dedicated escape chambers fore and

HMS Thetis *beached at Moelfre Bay in Anglesey after she sank on 3 June 1939 in Liverpool Bay with the loss of ninety-nine lives.* (Author)

aft. Yet in June 1939 ninety-nine men drowned in HMS *Thetis* in Liverpool Bay. As one observer recalled: 'Never will I forget the films and photographs of the stern of *Thetis* swinging in the tide of Liverpool Bay with men doomed to die so soon and yet so near to apparent help.'[6]

The cause of *Thetis*'s loss was the opening of the rear door of No. 5 torpedo tube while the bow cap was open – a dreadful combination of human error and mechanical and drill failure.[7] Despite the comprehensive escape arrangements, the men in *Thetis* died as a result of CO_2 poisoning on account of the overcrowding (there were 103 men on board) and the high level of activity.

During the Second World War escape had to take a back seat as submarines became more involved in the deadly business of combat. Escape hatches were, more often than not, bolted down from the outside to prevent their lifting under depth-charge attack. Twenty-one officers and men died in *P.32* in August 1941 while trying to escape from the engine-room. Unknown to them, the hatch had been secured from the outside so their efforts were in vain. Nevertheless there were some courageous escapes made by submariners of all nationalities. After the Second World War the Royal Navy's Ruck–Keene Committee began an international survey of escape experiences during the war, which found that nearly half the survivors had escaped without wearing any kind of breathing apparatus. On 6 October 1944 three officers escaped from *U168* without wearing any apparatus from a depth of 120ft. On 21 January 1945 Obersteurmann Klaussen escaped from *U1199* from a depth of 240ft. This remains the deepest 'real' escape (as opposed to a trial carried out by trained instructors). The Ruck–Keene Report was produced in 1946 and is perhaps the single most important submarine escape document ever written. Its conclusions were based on forty-six years of experience learned the hard way in various submarine accidents. The main conclusions of the Report, which still forms the basis for escape techniques for all modern submarines, were as follows.

The greatest hazard for the crew of a sunken submarine is the length of time it takes to flood up the compartment. Men should therefore be under pressure for the shortest possible period before escaping. In the case of HMS *Poseidon*, it had taken nearly 2½ hours to flood the compartment up, all the while the survivors were in great discomfort. Therefore the committee recommended the construction of one-man escape chambers from which the escaper could be ejected, by means of a piston, if he became unconscious.

The committee recommended the introduction of a new breathing apparatus to be used in the submarine prior to escape and discarded just before exit. This would enable survivors to cope with the effects of chlorine gas from the battery or a rapid build-up of CO_2 caused by overcrowding. Moreover, on the basis of war experience, the committee recommended that no breathing apparatus was required for the ascent itself. Members of the committee had observed American escape instructors at the 100ft tank in New London rising from the bottom without any apparatus, using a technique known as free ascent. In this method the body's own positive buoyancy is used to carry one upwards, while the excess gas arising from expansion in the lungs is exhaled through the mouth.

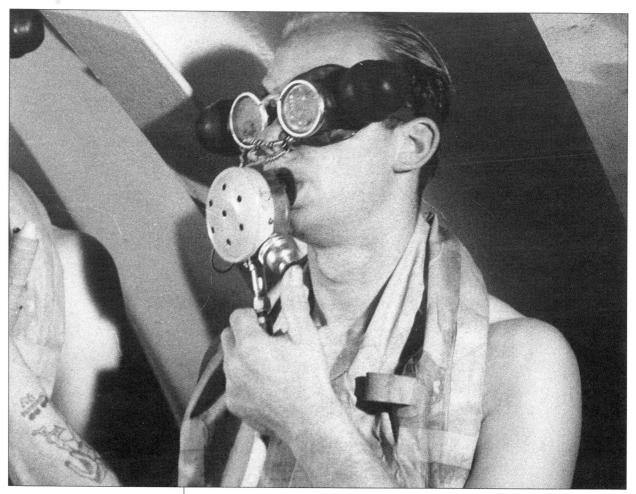

A British sailor breathes through his BIBS connection while waiting for the compartment to 'flood up' prior to an escape. Systems like BIBS relieved sailors of the stress of breathing air containing concentrated quantities of oxygen and carbon dioxide. (Author)

The most important recommendation was that efforts to escape from a sunken submarine should commence immediately. The survivors should not wait for rescue craft to arrive. This was a major factor in the loss of life in both *Thetis* in 1939 and *Untamed* in 1943. In both cases the survivors waited too long before beginning their escape. By the time they started the escape drill the levels of CO_2 in the boat were so high that the survivors were too weak or too confused to carry out the drill properly. This recommendation implied that survival on the surface after escape was of major importance. An immersion suit with a correctly shaped buoyancy life-jacket or stole was essential to maintain correct flotation of a possibly unconscious survivor and to provide thermal insulation. Tragic proof of this requirement came not long after the Report's publication and before this requirement could be implemented. On 12 January 1950 sixty-four men carried out flawless escapes from HMS *Truculent* which had been rammed by a merchant ship in the Thames Estuary. But only ten survived – the remainder drowned or died from exposure on the surface while waiting for rescue. Other recommendations included the phasing out of the Twill Trunk, improved air purification measures, location aids such as indicator buoys and smoke floats, and better training, including the construction of a 100ft escape tank similar to the American model.

PO L. Stokes, wearing a SEIS, prepares to enter the conning tower of HMS Tiptoe, 260ft down in the Mediterranean during submarine escape trials in 1962. (Author)

These extensive recommendations required a good deal of effort and it was some time before they could be incorporated into British submarine practice. Nevertheless since 1946 all the recommendations of the Ruck–Keene Report have been implemented and a Standing Committee on Submarine Escape (SCOSE) set up in 1951 (and still working) maintains a constant watch on policy and practice.

The years following the Ruck–Keene Report saw continuous development in escape techniques. Other navies have, to a greater or lesser degree, followed British practice and this remains one area of submarine activity, perhaps the only area, where international cooperation transcends the bounds of military secrecy. The first British submarines to be fitted with the dedicated one-man escape towers (which, however, lacked the piston to expel an unconscious man as required by Ruck–Keene) were fitted in the *Porpoise* and *Oberon* SSKs and to all British nuclear SSNs and SSBNs. British practice was adopted by the French Navy for their SSBNs and correlates closely with US Navy practice.

Along with increased depths from which an escape can be carried out came new techniques. Free ascent gave way to buoyant ascent, first used in 1953, in which the man wore a life-jacket or stole to provide the necessary buoyancy. The jacket was fitted with a valve which allowed the contained gas to expand during the ascent. This method ensured that all escapers who left a sunken vessel would reach the surface.

The disadvantage of buoyant ascent is that the expansion of gas is not constant but becomes progressively greater as one rises toward the surface. This means that the amount of gas which must be exhaled has to be carefully regulated; failure to achieve this leads to a situation where no more gas is available until further ascent, leading to further expansion. The solution lay in a system known as hooded ascent. The escape suit is fitted with a rubberised hood with a window at the front. The hood fills with air from the life-jacket exhaust valve which vents through a separate opening so that the air being breathed is constantly renewed during the ascent.

These developments have enabled escapes to be carried out from greater depths. In 1962 escapes were carried out from depths of 270ft from HMS *Tiptoe* and in 1965 from 500ft from HMS *Orpheus*. In the summer of 1970 escapes were carried out from a depth of 600ft from HMS *Osiris*. This recent account describes an escape from HMS *Oracle* at a depth of 300ft:

I climbed into the tower in the forward escape compartment of HMS *Oracle* and plugged into the air supply. I reported I was ready and the lower lid was shut below me. . . . Although not too late to turn back, I wondered why I had been foolish enough to volunteer and gave the signal to commence the flooding up of the tower. Almost immediately the thundering sound of water entering the tower commenced. There was no real increase in pressure initially but when pressure did start to increase it did so at a dramatic pace; indeed the pressure doubled every 5–6 seconds. . . . You can imagine my satisfaction as the pressure equalised. Then as I felt the compression on my chest I realised I had to breathe in. By this time the tower pressure was greater than fifty metres and the compressed gas was both hot and viscous and I felt the hot gas burning as it entered my lungs.

To my great relief I felt the upper hatch open above my head and I could not prevent myself from rising out of the tower due to my considerable buoyancy. The open hatch brushed against my back as I left the tower (the submarine was under way) and so I commenced my ascent. I was temporarily dazzled by a bright light outside the tower (illumination for a video camera) before being plunged into the darkness of a Scottish loch 300ft below the surface. As the water rushed past me the colour gradually returned. Starting as black, passing through dark green and then blue, it seemed an eternity as the blue colour became brighter and brighter and then suddenly there was the blinding white light of the sun which had come out just a few moments before. A maximum speed of ascent of just over three metres per second, reached just prior to surfacing, meant that as I broke the surface, all but my feet bobbed out of the water like a cork. I then fell backwards giving the 'thumbs up' signal triumphantly to indicate that all was well. The whole ascent had taken only thirty seconds.[8]

Other escapes using the tower method have taken place from depths as great as 600ft,[9] but this is reckoned to be the greatest depth from which such an escape is possible. The tower method does offer a real chance of a crew escaping from a sunken submarine without outside assistance.

The introduction of the deep-diving nuclear-powered submarine in the early 1960s posed fresh problems for those charged with the development of escape techniques. Submarine test depths now exceeded the limits of escape systems then in use. The US Navy's McCann Rescue Bell was considered effective down to depths of 850ft (under ideal conditions) but even this depth was exceeded by the new fleet submarines coming into service. The loss of the American submarine *Thresher* in 1963 (as a result of a reactor failure which caused the loss of all power on board) cruelly exposed this problem. The Deep Submergence Rescue Vehicle (DSRV) was designed to remedy this deficiency.

The DSRV is a very small submarine of only 37 tons displacement but reportedly capable of diving to depths as great as 5,000ft – far in excess of test or collapse depths. It can be loaded on to US Air Force C-141 Starlifter or C-5A Galaxy transport aircraft for rapid transport to the scene of an accident. Originally twelve such craft were requested by the USN but ultimately only two were completed: *Mystic* and *Avalon*. They completed in 1971–2 and can carry twenty survivors. Today one is stationed on the US Atlantic coast and the other on the Pacific coast.

The DSRV is carried to the scene of an accident on the casing of another submarine. It is operated by being positioned over the escape hatch of the submarine in trouble. It then extends a 'skirt' over the hatch making a watertight seal. Those inside the submarine can then release the hatch. The DSRV can then embark a number of survivors, take them to the host submarine and then return for the next group. In a rare example of super-power cooperation the DSRV's skirt is compatible with the escape hatches on American, British, French and Russian submarines. In 1979 a joint Anglo-American exercise tested the concept. HMS *Odin*, a British SSK,

CC Henri Rousellot shuts the upper hatch on Rubis's *bridge before going below. Attention to basic details of submarine safety – like not leaving hatches open or unattended is a pre-requisite of submarine safety. (ECPA)*

'sank' in 400ft of water off the Isle of Arran. A DSRV from San Diego in California was flown to Glasgow's Prestwick airport and then sent by road to the British submarine base at Faslane on the Clyde. There the DSRV was 'mated' to the British SSBN *Repulse* which then steamed out to the exercise area. The DSRV was released in 250ft of water and successfully recovered a token number of *Odin*'s crew. The time taken from the alert being raised at San Diego to embarkation on *Repulse* at Faslane was about forty-seven hours. Recently a DSRV was successfully mated with the British Trident SSBN *Victorious* to ensure compatibility.

The Royal Navy has not produced any dedicated rescue craft of its own. Instead Britain has chosen to exploit long-standing links with the USN to access use of American DSRVs. However the Royal Navy has carried out trials with a number of commercial submersibles used by the extensive oil and gas industry in the North Sea. Presumably in the event of an accident in UK waters one of these craft could be brought on to the scene fairly quickly. American DSRV practice has been copied by Sweden which maintains its own URF submarine rescue vehicle. This can be transported by road to any point on the Swedish coast in the case of a submarine accident. The Russian Navy has not only constructed its own DSRVs but also a parent submarine to carry them. Two India-class SSKs have been built, one serving with the Northern Fleet, the other with the Pacific Fleet. They are unarmed but carry two DSRVs in tandem wells fore and aft of the conning tower. They have the additional advantage of being available for special operations in times of international tension or warfare.

One problem associated with DSRV-aided escapes is that the submarine in distress somehow has to reveal her position. All navies have procedures, such as the British *Submiss* and *Subsunk*, for dealing with submarines which have failed to make routine surfacing reports. Even so, as in the case of the *Affray*, it was some time before the submarine was located. This problem was highlighted in May 1968 when USS *Scorpion* sank off the Azores (after an explosion caused by a torpedo 'running hot' in the tube). It is vital that the submarine is able to give an accurate indication of her position. In the past they were equipped with signal buoys for this purpose but there is a natural fear that the buoys will be noisy and that they could become dislodged under operational or combat conditions. During the Second World War British and American submarines had their signal buoys welded down for fear that they would break loose in a depth-charge attack.

Other modern escape options include fitting the submarine with an escape chamber that is integral with the hull and can be released in an emergency – a form of 'non-returnable' Gerolami-Arata-Olivati system described above. German U-boats of the postwar *Bundesmarine* are fitted with such 'Gabler Spheres' named after Prof. Ulrich Gabler, the designer. The main theatre of operations for the *Bundesmarine* is the Baltic, a comparatively shallow sea, where such an escape would be practicable. The Russian Navy has also fitted such escape compartments to its submarines.

For as long as submarines are in service there will be a requirement for escape and rescue equipment. However, the only truly safe method of escape is *not* to have an accident.

NOTES

1 Shelford, W.O., *Subsunk!* (London, Harrap, 1960), p. 21.

2 For the story of the rescue of those in *Squalus* see Shelford op. cit., pp. 70–81.

3 HMS *Poseidon* was rammed and sunk on 9 June 1931 in the Yellow Sea. Six men escaped from the fore ends.

4 Quoted in Shelford, op. cit., p. 92.

5 PRO ADM1/9373 minute by Director of Plans.

6 Tabb, H.J., 'Escape from Submarines. A Short Historical Review of Policy and Equipment in the Royal Navy', *Transactions of the Royal Institute of Naval Architects*, 1974, p. 31.

7 The most comprehensive account of the loss of *Thetis* is still Warren, C.E.T. and Benson, J., *The Admiralty Regrets . . .* (London, Harrap, 1958).

8 Turner, Surg. Lt M., 'Submarine Escape from a Depth of 300 ft – A Personal Experience', *Journal of the Royal Naval Medical Service*, 1995, vol. 81, pp. 212–13.

9 Donald K.W., 'A Review of Submarine Escape Trials from 1945 to 1970 with a particular emphasis on Decompression Sickness', *Journal of the Royal Naval Medical Service*, 1991, vol. 77, pp. 171–200.

FRIENDLY FIRE

'Don't shoot! Steve here!'
Signal sent by the commanding officer of HMS *Turpin* after an
encounter with HMS *Tapir*, April 1945

riendly Fire – the act of firing on one's own forces – is as old as the history of war. However, in terms of naval warfare it is a comparatively recent phenomenon. In the days of sail, actions were fought at slow speed and at very close range. There was plenty of time to identify an opponent before the action began. The technological revolution in the late nineteenth and early twentieth centuries transformed naval warfare. Engagements would now be fought at greater ranges and at greater speeds. The introduction of the aircraft and the submarine transformed the naval 'battlefield' into a three-dimensional one with engagements being conducted on, above and under the ocean.

Submarines were, and are, particularly vulnerable to the attentions of their own side. Since 1914, twenty-four submarines have fallen victim to attacks from forces of their own side (with another two unconfirmed losses). By far the worst culprits are the British and Americans. This is not to be taken as an indication of the inadequacy of Anglo-American command methods but reflects on the fact that Britain and America were the only two powers engaging in large-scale, offensive, airborne anti-submarine operations in waters used by their own submarines. On the other side, Axis airborne ASW was limited to their own littoral waters. Mistakes were bound to happen.

There are four reasons for these 'own goals' as some authorities describe them. First, there is a simple failure of recognition – a case of mistaking Brown for Braun. Second, errors in navigation may cause units to be out of position. Third, there is the failure of measures designed to separate friendly forces. A fourth, related, cause is the failure of commands to keep all units informed of the disposition of friendly forces.

All submarines, regardless of nationality, are remarkably similar in appearance. In anything other than a flat calm only the conning tower is usually visible above the waves and in the excitement of battle it is very easy to assume that the submarine in question is hostile. Submarine recognition and identification are hard enough on a clear day and with good visibility. In bad weather, darkness or through a periscope it is far harder. In the middle of October 1942 the Japanese submarine *I.25* (Lt Cdr M. Tagami IJN) arrived off the coast of Oregon on the west coast of the United States for a patrol which was to include two raids on local forests by the submarine's light aircraft armed with incendiaries. By the middle of October she was on her way back to Japan with only one torpedo left. On the morning of 11 October one of her look-outs reported masts on the horizon. *I.25* dived and Tagami identified the ships as two submarines proceeding on the surface. Understandably perhaps, Tagami assumed that the submarines were American and fired his remaining torpedo which scored a hit, sinking one of the submarines. Unfortunately they were not American but Soviet. *L.15* and *L.16* were proceeding from Vladivostock to Murmansk via Panama but when *L.15* arrived at San Francisco on 15 October she reported that *L.16* had sunk after an internal explosion on 11 October. This explanation was accepted until the American postwar investigators were able to analyse Japanese operational papers – only then did the true fate of *L.16* became known. Although Russia was at war with Germany at the time, she remained scrupulously neutral as regards Japan since Stalin did not want to be engaged on two fronts simultaneously. However, the diplomatic consequences of Tagami's failure to identify his target can be imagined.

The tendency to assume that a submarine is hostile is particularly true of aircrew. A crew could go for months without seeing so much as the 'feather' thrown up by a periscope. Suddenly a submarine would appear and in the excitement of the moment, recognition drills were sometimes quickly forgotten. As one British submariner, Lt John Coote, the 'Third Hand' of HMS *Untiring* put it: 'All the restrictions in the world didn't guarantee immunity from a keyed-up Wellington pilot dropping out of the overcast and seeing his first ever U-boat right in front of him with Distinguished Service Order written all over it.'[1] This was particularly true of attacks carried out in waters used by submarines of both sides. During the Second World War British submarines would transit the Bay of Biscay while proceeding to and from the Mediterranean. The same waters were extensively patrolled by the RAF looking for U-boats heading to and from the Atlantic. Cdr Arthur Pitt of HMS *Taku* commented: 'Coming back to the UK from the Mediterranean was a nightmare. Although we were protected by a moving haven I was never confident about the ability of our own aircraft to respect it. The people who really frightened me were those bloody Poles. Even if you painted the boat red, white and blue they would still shoot first and challenge later. Submarine sanctuaries were a contradiction in terms to them.'[2]

The fitting of radar to submarines improved the chances of identifying and recognising a target. In April 1945 radar prevented a potential disaster when two British submarines, HMS *Tapir* (Lt Cdr J.C.Y. Roxburgh RN) and *Turpin* (Lt Cdr J.S. Stevens RN), encountered each other off the Norwegian coast. Both submarines had recently been fitted with the new Type 267W surface warning radar set which gave much better coverage than the Type 291 set previously fitted. *Tapir* and *Turpin* went out on patrol together in April 1945, occupying adjacent billets off the Norwegian coast, and great things were expected from the new radar set. As it turned out, 267W proved to be *Turpin*'s salvation.

On the night of 6 April 1945 *Tapir* picked up the HE of another submarine. *Tapir*'s commanding officer, now Vice Adm Sir John Roxburgh, recalled:

We heard him surface so I immediately broke surface to get my 267 on it. I got a lovely blip on it so I tracked it very quickly with my standards sticking up about 5 or 6 feet to get a picture.[3] He was going away at speed so I surfaced to chase him and was about to overtake him, but he was about to leave my area and go into Stevens' who was patrolling next door. I wasn't going to catch up with him so I got ready to send an enemy report when suddenly to my astonishment I got the correct challenge from the other side immediately followed by 'Don't shoot, Steve here!'[4]

Turpin had found herself out of position and had surfaced in order to proceed to her area at speed. Her 267W operator had detected *Tapir* coming up astern so Stevens reduced speed to assist the development of a range and bearing plot. It was only when *Turpin*'s 267W operator reported mutual interference from a similar 267W set that Stevens realised that the other submarine could only be *Tapir*. On this occasion the correct procedures worked. *Tapir* did not go out of her area and the correct challenge and reply were given.

A second reason for attacks on submarines is navigational error. Without a sight of land or a reliable solar or astronomical observation, most navigation was done by dead reckoning, making allowances for wind and current. On 2 May 1942 the Polish submarine ORP *Jastrazb* (Lt Cdr B.S. Romanowski) was sunk by the Norwegian destroyer *St Albans* (Cdr S.V. Storeheil) and HMS *Seagull* (Lt Cdr J.Pollock RN). *Jastrazb* was formally commissioned into the Polish Navy in November 1941 at the submarine base at New London, Connecticut. After a refit at Blyth she sailed for her first and only patrol on 22 April 1942 in an area 180 miles north-west of Altenfjord in northern Norway.

Romanowski and his crew endured a week of bad weather and mechanical difficulties. Then, at 16.00 hrs on 2 May, he sighted a U-boat running on the surface and prepared to attack. His attempt, however, was plagued with mechanical problems, and Romanowski reluctantly had to abandon the attack. At about 19.45 *Jastrazb* picked up the noise of approaching engines on her hydrophones and after a while Romanowski could see the ships through his periscope. After a quick look at his recognition book, he identified them as friendly and ordered the release of smoke candles to avoid being attacked. The smoke candles were fired from a 3in underwater 'gun' built into the submarine's pressure hull and ignited on reaching the surface. However, the signals were not seen by the two approaching ships which delivered a devastatingly accurate depth-charge attack. The submarine's pressure hull was fractured in a number of places and water poured into the hull, contaminating the batteries which began to give out chlorine gas. Romanowski had no choice but to bring *Jastrazb* to the surface.

Jastrazb's assailants were the Norwegian destroyer *St Albans* and the minesweeper HMS *Seagull*: both were part of the escort for convoy PQ.15 heading for Murmansk. They were steaming on the convoy's port bow – the opposite side of the convoy to where an Allied submarine might be expected – when they gained a solid Asdic contact and attacked it with depth-charges in position 73° 30'N, 17° 35'E. The smoke candles fired by

Jastrazb were not seen by either ship before the depth-charges were fired. *Jastrazb* surfaced between the two ships but with her bows pointed toward *St Albans* so that her pendant number, *P.551*, was not visible. Both ships opened fire on the submarine with machine-guns and *St Albans* also brought her 3in gun into action. Romanowski had been first on to the conning tower, followed by his British liaison officer and two Royal Navy signal ratings. The two ratings were killed instantly, along with three Polish ratings, and six more crew were wounded, including Romanowski, before *Seagull* recognised the pendant number and both ships ceased fire. Boats were launched to offer assistance but it was clear that *Jastrazb* was beyond salvage. The survivors were taken on board HMS *Seagull* (Romanowski even managed to bring all his Confidential Books with him – no mean feat!) before being transferred to the cruiser HMS *London*.

What had gone wrong? In theory the course of PQ.15 should have taken the convoy well to the north of the submarine's patrol area. However, owing to heavy ice, the convoy had altered course to the south at 09.45 hrs that morning and was steaming diagonally across the submarine's intended track. On the other hand *Jastrazb* was hopelessly out of position, by as much as 100 miles, to the north-east of her area. At the Inquiry into her loss, held in HMS *Forth*, her navigating officer said that he had been unable to take a sight of the sun or stars for five days, and had been navigating entirely by dead reckoning. His notes were examined and it was found that the allowances he had made for currents and tidal sets were accurate but that it was practically impossible for him to navigate properly since *Jastrazb* was not fitted with a bottom log nor was there any accurate method of checking engine revolutions. Under these circumstances an error of nearly 100 miles was understandable. The inquiry concluded that *Jastrazb*'s loss was a misfortune of war, compounded by the difficulties of navigation in such high latitudes, and exonerated all three commanding officers.[5]

A third reason is failure of identification and deconfliction procedures. These are often associated with navigational errors. Identification procedures exist so that one vessel can identify herself or check the identity of another vessel. They consist firstly of physical markings painted on the submarine to aid visual recognition; secondly, the firing of coloured flares, grenades and rockets, and finally the use of the coded challenge/reply system. This comprised three-letter codes which did not make up a recognisable word. Thus TYB would be the 'challenge' and WVP the 'reply'. The message would usually be sent by directional light. For security reasons the letter groups were changed at regular intervals.

Yet even these comparatively simple methods could fail. On 12 March 1918 submarine *D.3* was on anti-submarine patrol in the English Channel off Dieppe when she was sighted by the French airship *AT-0*. *D.3* attempted to identify herself using rocket signals but the crew of the airship thought that the rockets were being fired at their vulnerable hydrogen-filled balloon. Moreover, the rockets were not the recognised inter-Allied recognition signal. Accordingly, they returned fire with a machine-gun and the submarine dived. The airship crew immediately (and reasonably) assumed that the submarine was German and dropped six 52kg depth-bombs. Amid the explosions *D.3* came to surface briefly before sinking back into the depths. Four survivors were left swimming in the water, one of whom yelled to the airship crew as they hovered overhead, 'You got us!'. (The true response was probably a good deal stronger.) At this point Lt Saint Remy, *AT-0*'s commander, realised that they might be British. Sadly, he could not pick them up and while he was searching for a ship to rescue them, the men drowned. The Frenchmen did not see the deck recognition mark, a white circle on a black background, painted on *D.3*'s fore hatch.

Submarine commanders were understandably reluctant to place their trust in such recognition procedures. There were a number of instances where submarines were attacked despite having identified themselves. Submarine commanders faced a dilemma: either identify yourself and risk being attacked or dive into safety and hope to get away with it. In July 1942 HMS *Talisman* (Lt Cdr M. Willmott RN) was on the surface when her look-outs spotted a Sunderland flying-boat. Willmott assumed that the aircraft had not seen his submarine and decided that the safest thing to do was to dive quietly out of sight without making a recognition signal. Alas, the *Talisman* had been spotted. Her quick dive was interpreted by the Sunderland's crew as 'an admission of guilt' and she was attacked. The depth-charge explosions forced *Talisman* down to a depth of 355ft – well below her maximum diving depth of 300ft. Drastic action was required to halt the dive yet Willmott could not risk disclosing the boat's position through too violent blowing of the ballast tanks for fear of a further attack. Instead he managed to arrest the initial deep dive at 200ft and then allowed the boat to increase depth gradually while at the same time altering course. In his comments on the incident Flag Officer Submarines Vice Adm Max Horton gave credit to Willmott whose 'experience and cool handling saved what might have been a disaster, though he was himself to blame for the incident which might have caused this disaster.'[6]

In retrospect Willmott would have done better to have identified himself to the aircraft before diving, given the good visibility and the fact that he was not in a submarine sanctuary. However, his caution, in view of the number of attacks on submarines by 'friendly' aircraft, is understandable.

Care had to be taken when giving or replying to a challenge that the correct group was used. In February 1941 the British submarine HMS *Thunderbolt* (Lt Cdr C.B. Crouch RN) was ordered to Halifax to join the North Atlantic Escort Force. On 24 February she was heading westwards after parting company with her escort, the corvette HMS *Asphodel*. Lt John Stevens was her first lieutenant:

A week after we sailed, I was OOW on the bridge for the morning watch. The weather and visibility were good, with a moderate swell and low grey cloud. Searching the horizon through powerful binoculars, I sighted the mast and funnel of a hull-down ship. I ordered 'diving stations' and called for the captain to come up to the bridge. Our courses were now converging, and as she drew nearer, she was identified as a large grey-painted passenger ship. As there had been no information about any friendly forces in the area we assumed her to be a German or Italian vessel. The torpedo tubes were brought to the ready and when we were three miles away from her Crouch ordered the

Opposite: HMS Tribune *comes along the depot ship* Forth *in Holy Loch in 1940. She is flying two White Ensigns as an indicator of her nationality to supposedly 'friendly' aircraft. (Author)*

Some of the crew of the British submarine Talisman, *attacked by a Sunderland flying boat and forced down to a depth of 300ft.*
(US Navy)

HMS Thunderbolt *was the raised and re-commissioned* Thetis. *In February 1941 she came under attack from an armed British merchant cruiser and was lucky to escape unscathed.*

signalman to make the challenge. The signalman spelt out the group of letters using a large Aldis lamp but there was no reply. Instead she swung round under full helm and opened fire on us with her main battery. The shell splashes came alarmingly close and Crouch cleared the bridge and pressed the diving klaxon.[7]

The ship was the British armed merchant cruiser *Canton* and her fire was accurate: *Thunderbolt* was lucky to get away unscathed. But why had *Canton* opened fire when the submarine had given the challenge? The problem was that *Thunderbolt* had not made the challenge then in force: '. . . the problem with these long Atlantic voyages was that routine relaxed and everyone got a little careless. The signalman hadn't checked the correct challenge for the day, which is why we were fired on.'[8]

Deconfliction is the intentional separation of friendly forces. Submarines operated independently inside designated patrol areas, known as 'billets' in British parlance. They were not supposed to move out of their areas unless in the most serious circumstances. The rationale behind these precautions was not only to protect the submarines from the attentions of their own kind, but so that if another submarine were sighted the commanding officer had complete freedom of action. As Cdre Howard Kelly trenchantly reported when commenting on the circumstances surrounding the loss of the Italian submarine *H-5* on 16 April 1918: 'When stationed in definite zones, any submarine sighted in that zone must be attacked without hesitation or consideration; there can be no question of recognition.'[9]

These precautions could be double-edged. In 1942 HMS *Upright* was on patrol on the surface in the Mediterranean in the early hours of the morning when her look-outs sighted another submarine, also on the surface and on a reciprocal course. Owing to the large number of British submarines at sea at the time, there was an embargo on night attacks on other submarines because of the difficulty of correct identification. The two vessels passed at about five cables' distance and when the other submarine was slightly abaft *Upright*'s beam she was identified as a large U-boat. Capt George 'Shrimp' Simpson, CO of the 10th Submarine Flotilla to which *Upright* belonged, commented, 'another opportunity of destroying the enemy lost through the unfortunate but indispensable necessity of enforcing a submarine "armistice" when more than one submarine is operating in the same area'.[10] It is not recorded whether the U-boat, whose identity is sadly unknown, was operating under a similar embargo, or whether her look-outs simply failed to spot the much smaller British submarine.

During the First World War British, French and Italian submarines maintained a patrol off the Austrian submarine base of Cattaro (now Kotor) in the Adriatic. During the day this was a dived patrol with the boats deployed close inshore. At night the boats retired seaward to surface and recharge their batteries. Each submarine was allocated an area out of which she was not supposed to go. On the night of 17 April 1918 the British *H.1* (Lt John Owen) sighted another submarine in her area. According to the proceedings of the subsequent Board of Inquiry:

The captain of *H.1* sighted at 7.45 p.m. a submarine on the surface. He supposed that it was an enemy submarine as no Allied submarine ought to have been in the place where he supposed himself to be. He asked his navigating officer, Lt Hayes, to look through the periscope and say whether it was an H-class submarine. The navigating officer replied that it certainly was an enemy submarine. The captain said that it looked very much like a UB type which has a flush upper deck and no raised forecastle. Lt Owen continued to attack and fired two torpedoes one of which hit the submarine on the port side.[11]

Owen and his navigator were both wrong. The submarine was the Italian *H.5*, hopelessly out of position and virtually in the middle of *H.1*'s area. Moreover, she had surfaced before sunset, contrary to operational practice. In so doing, according to Cdre Howard Kelly, the British commodore at Brindisi, she was 'looking for trouble'.[12] There were only four survivors. The subsequent inquiry absolved *H.1* of any blame but it was decided that in future the practice of sending British, French and Italian boats to sea together would cease. Henceforth the submarines of one nation would provide the patrol at any one time. Fixed patrol areas such as those in which *H.1* and *H.5* were operating are designed to allow for optimum deployment of available assets and to protect submarines from the over-zealous attentions of their own side. *H.5* had drifted out of her area by accident but in October 1940 an Italian submarine was sunk in circumstances that bordered on the criminal. The three submarines *Gemma*, *Tricheco* and *Ametista* were ordered to set up a patrol line off the Caso Channel at the eastern end of Crete. The submarines were disposed to the north, south and middle of the channel to catch any British vessels landing troops and supplies on Crete. On 7 October 1940 the *Tricheco* was given permission to finish her patrol and return to base at Leros. In doing so she passed through the patrol area occupied by *Gemma*. When *Tricheco* sighted another submarine on the surface at 01.21 hrs on 8 October her commander did not consider the possibility that the submarine might be *Gemma* occupying her rightful patrol area. *Tricheco* fired two torpedoes and saw both of them hit. There were no survivors from *Gemma*. This was a dreadful error of sheer carelessness by *Tricheco*'s commanding officer. *Gemma* had been sunk for two reasons: first, because *Tricheco*'s commanding officer did not make himself familiar with the disposition of friendly forces in the area and second, because he made no attempt to confirm the identity of the target before firing.

While *en route* to and from their patrol area, submarines were protected by a 'sanctuary' or 'moving haven'. This was an area of ocean corresponding to the submarine's track, and its position would be adjusted daily to correspond with the submarine's progress. For additional safety the Total Bombing Restriction Area was extended 20 miles either side of the submarine's track, and a further 20 miles each side of the track would be added to compensate for navigational errors by the aircrew, making a total of 40 miles either side of track. If there was a chance of two submarines passing through a sanctuary simultaneously this information would be relayed to the boats concerned well in advance. Attacks on any submarines were forbidden in these sanctuaries. If a 'genuine' enemy

*Lt-Cdr Hugh Mackenzie,
commander of* Thrasher.

submarine was sighted it could not be attacked but had to be reported by radio. Obviously, for these precautions to work accurate navigation was essential.

On the evening of 26 July 1942 HMS *Thrasher* (Lt Cdr H. Mackenzie) sailed from Port Said for patrol. Mackenzie later recalled that they were:

. . . proceeding on the surface, zigzagging with a safety corridor around us to prevent attack by friendly aircraft. At about 9 p.m. I came down from the bridge to have supper thinking that everything was all settled, we were about thirty miles outside Port Said then, happily on our way. I was sitting at my usual place at the wardroom table, which was in the gangway as near to the control-room as possible, so that I could get out in a hurry if needed, and I remember very clearly something telling me to get out on the bridge and I got up and left the table without explaining why, or in response to any request for me to go on the bridge. I was just going through the watertight bulkhead between the wardroom and the control-room when there was the most God-awful explosion and the next thing I remember was picking myself up from the floor at the after end of the control-room up against the W/T office. During the course of getting up from there I remember seeing the ship's galley range – which was in the passageway abaft the wireless office – leave the ship's side in a cloud of blue sparks.

As I was picking myself up I heard the OOW, the first lieutenant, giving the order to dive and the drill was carried out. The vents were opened by the outside ERA although the lights had gone out and there were a lot of odd noises going on. I then heard the Bren gun on the bridge open fire so I knew they were all right up there. The signalman, look-outs and OOW came down and we dived. Reports were coming in of damage: quite clearly we had no electrical power, smoke was coming from the battery compartments, we had a fire in the battery (smoke and gas were coming from Nos 2 & 3 and No. 1 battery was on fire and later had to be flooded with distilled water) and we had to drop all the main fuses, but we sorted ourselves out. The gyro was humming like an angry bee and the magnetic compass was upside down and no good. It seemed to me that the best thing we could do would be to get back on the surface, especially as the first lieutenant thought it was a Swordfish which had attacked us.

It was indeed a Swordfish. The attack had been carried out at 21.40 hrs in position 31°N 32°E, in the middle of a submarine 'sanctuary' where total bombing restriction was in force. The Swordfish, from 815 Sqn (a Fleet Air Arm squadron) and flown by Sub Lt D. Stuart RNVR was on anti-submarine patrol and was armed with four 250lb depth-charges. The aircrew were not advised of the sanctuary when they took off from El Gamil airfield and all attempts to contact the plane after she had taken off were unsuccessful. Ken Sunor was a Telegraphist Air Gunner (TAG) with 815 Sqn and at the time was waiting at the airfield to take off on the next patrol. He recalled: 'I was due to take off on the next patrol so was sat in the W/T van when our duty officer arrived – in some concern. Could the Tel [Telegraphist] raise the aircraft in the air and send an NYKO message? It warned him of the *Thrasher* – information only just received. But as we called, so the aircraft sent us his message that he had attacked. It was too late.'[13]

The Swordfish dropped all four charges: one fell on the port side abreast the fore hatch and exploded under the ship, two others fell on the starboard side abreast the after end of the midship's external torpedo tubes while the last was not seen to explode. The force of the three explosions lifted the stern so much that *Thrasher*'s bows were driven under water. Now Mackenzie had to bring his crippled boat back to harbour:

It was quite clear that we could not proceed with the patrol so we surfaced and then had to try to get our engines started with no electrical power which took a little time. . . . We got the engines going but our main problem was to keep afloat because the main ballast tanks were leaking the whole time and we had to keep blowing them with high-pressure air since there was no electric power to use the LP blower. We'd got no means of steering a proper course since all the compasses were gone. Fortunately it was a starlit night and I knew that Port Said was roughly south so I directed operations standing on the bridge with the Pole Star behind me saying 'Port' and 'Starboard' down the voicepipe.

During an attack from a Fleet Air Arm Swordfish aircraft in the Mediterranean, HMS Thrasher's *stern was lifted into the air and her bows driven underwater by the force of three exploding depth-charges.* Thrasher *sustained severe damage.*

We then ran into one of the trawlers patrolling off Port Said and identified ourselves. We staggered into Port Said on our last gasp of high-pressure air . . . our ballast tanks were awash.[14]

Captain *S1*, Captain S.M. Raw RN, commented that *Thrasher*'s 'excellent construction was very severely tested and proved thoroughly sound'.[15]

The RAF, under whose control the Swordfish was operating, launched an inquiry into the incident under the auspices of 201 Group. Ken Sunor recalled: 'There was of course an inquiry. The aircrew said that it [the submarine] was outside the sanctuary. The *Thrasher* of course said they were inside. At that distance from the coast and with radar I don't think the aircrew were much in error – it was probably a borderline case – but a more mature crew might have given the submarine the benefit of the doubt.'[16]

The sinking of the French submarine *La Perle* represented a complete failure of both identification and deconfliction procedures. On 8 July 1944 *La Perle* was returning to the Mediterranean after a refit in Philadelphia Navy Yard when she was bombed and sunk by Swordfish aircraft operating from the MAC ships *Empire Maccoll* and *Empire Maccallum* which were part of the escort for convoy ONM.243. There was one survivor, CPO Emile Cloarec, who was picked up by HMCS *Hespeler*. He reported that fifteen of *La Perle*'s ship's company of fifty-eight officers and men had escaped from the submarine before she sank but the others had been unable to keep afloat.

That the attack should have occurred at all was a cause for concern. *La Perle*, under the command of CC Tachin, left New London Conn., on 26 June for St John's Newfoundland under escort by the American destroyer *Cockerel*. After a short stay in Newfoundland she sailed for Holy Loch in Scotland. In coastal waters she was escorted by the Canadian destroyer *Chicitoumi* but would make the crossing of the Atlantic alone and travelling on the surface. Sailing across the Atlantic at the same time and on a roughly similar course was the Halifax–Clyde convoy ONM.243 which included the MAC ships *Empire Maccoll* and *Empire Maccallum*. The convoy was escorted by the C.5 escort group commanded by acting Cdr C.H. Stephen OBE, DSC, RCNR, in the destroyer HMCS *Dunver*. Since sailing, the escort group had received daily situation reports from Western Approaches headquarters at Liverpool which included details of *La Perle*'s movements, but signals advising friendly forces of the bombing restrictions in force around *La Perle*'s likely position were not passed to the escort group. However, adequate information was available to Cdr Stephen to indicate that *La Perle* would pass sufficiently close to the convoy to be within the area covered by his air patrols. He realised this and sent two signals on 7 July warning of *La Perle*'s proximity to the convoy. The signals were to be passed by HMCS *Dunver* to the convoy commodore, vice-commodore and the MAC ships, but the correct procedure was not carried out and the signals were not received in the MAC ships. As no evidence was available from the commodore's ship it is not possible to establish why the procedure was not adhered to.

Prior to the convoy's departure, a general agreement with regard to air patrols was made, by telephone, between Cdr Stephen and the air staff officer in *Empire Maccoll*, Lt Cdr Neale. No patrol orders were given to the MAC ships while at sea, so no special precautions were taken to advise aircrew prior to morning patrols on 8 July that a friendly submarine was in the area. Stephen may also have been lulled into a false sense of security by a message received at 00.38Z on 8 July from Western Approaches headquarters, which placed the submarine, wrongly, further away from the convoy than she actually was.

La Perle was first sighted by a Swordfish at 12.53Z. The pilot, Lt Otterveanger of the Royal Netherlands Navy, resolved to shadow the submarine and call up reinforcements rather than make an immediate attack which he felt might not be successful, given the quick diving time of a U-boat. He noticed the recognition signals made by *La Perle* but disregarded them. Between the time of *La Perle*'s sighting and the attack being carried out, an interval of more than an hour elapsed. Surprisingly, neither the pilot nor the air staff in *Empire Maccoll* thought it strange that the supposed U-boat should remain on the surface, keeping a steady course and doing an estimated 15 knots, and making no attempt to dive into safety. It was not until 13.58Z that Stephen realised that the submarine his aircraft were chasing might be *La Perle*. Even then, there was no sense of urgency about his signal and no attempt was made to halt the attack by communicating directly with the aircraft. Stephen's failure to realise the situation was probably influenced by the latest Admiralty intelligence report indicating that a U-boat might be in the vicinity of the convoy.

Before take-off, the aircraft were advised of the current recognition signals then in force. On sighting the aircraft, *La Perle* made the correct signals in good faith, having been informed of the total bombing and attack restrictions in force around her – but her signals were totally disregarded by the aircraft. Presumably because the pilots had not been warned of *La Perle*'s presence, they disregarded any signals coming from a potentially hostile submarine. Once Otterveanger had been joined by the other seven Swordfish, he led the attack, dropping three depth-charges alongside the submarine. The explosions stove in *La Perle*'s hull in the region of the control-room causing flooding which in turn caused electrical fires. CPO Emil Cloarec had just asked permission to spend a quarter of an hour on the bridge when the attack began.

Directly beneath the bridge in French submarines was a small compartment, known as the 'kiosque', containing the attack instruments. It was in this compartment that Cloarec was standing since the bridge was occupied by all five of the submarine's officers and a quartermaster. The fire in the control-room vented up through the conning tower and most of the men there and the officers on the bridge were horribly burned. Lt Zappert, the Royal Navy liaison officer, fired off a number of Very cartridges indicating that the submarine was 'friendly' but to no avail.

On receiving reports that the fire and flooding were out of control, CC Tachin gave the order to abandon ship. *La Perle* began to settle by the stern and eventually sank twelve minutes after the Swordfish attack. Cloarec, together with fourteen other members of the crew, had escaped from the submarine and remained swimming. One by one, the Frenchmen drowned or succumbed to exposure until only Cloarec was left alive. He was eventually picked up by the Canadian destroyer *Hespeler* which had been detached by Stephen to look for survivors. He was practically unconscious and initially taken for a German seaman. It was only when he was heard to speak French, *Hespeler* having a number of French Canadians in her ship's company, that the awful truth of what had happened was confirmed.[17]

The French Navy received an expression of regret for what happened but it was not enough. The French demanded a full-scale inquiry. This was held under the chairmanship of Rear Adm Lionel Murray CBE, Commander-in-Chief Canadian North West Atlantic, at St John's, Newfoundland. If the French wanted blame to be apportioned then they were to be disappointed. Cdr Stephen was exonerated, as were the aircrew from the MAC ships. Only the signals officer in HMCS *Dunver*, Lt Benson, was reprimanded. The sad affair of *La Perle* is fraught with questions. Why was *La Perle* given a route that would take her so close to ONM.243? Why were Stephen's two signals not received in the MAC ships? Why did the aircrew ignore the correct recognition signal when it was given by *La Perle*? Most important of all, why did *La Perle* not dive, rather than bother with identification, as soon as the Swordfish was sighted at 12.53 hrs?

Faulty navigation by both aircraft and submarine were responsible for the loss of the Soviet submarine *V.1* (ex-British HMS *Sunfish* under the command of Captain 3rd Rank Israel Fisanovitch) in July 1944. The submarine was *en route* for Murmansk from Rosyth but failed to arrive. A report was subsequently received from an RAF Liberator that it had attacked and sunk a submarine in the Norwegian Sea on 27 July 1944.

Rear Adm Egerton, the SBNO (Senior British Naval Officer) North Russia, considered that the submarine sunk by the Liberator could have been *V.1*. His initial impression was confirmed when *V.2* (ex-HMS *Unbroken*) arrived on 3 August. Egerton interviewed Leading

Telegraphist C.A. Wilkes, one of the British liaison party on board the submarine, who told him that the Soviet commanding officer had not kept to the 'safe' route allocated to him and that recognition flares were never kept on the bridge where they were immediately accessible. Egerton concluded: 'It seems more than possible therefore that the submarine attacked on 27 July was in fact *V.1* who was off her route and who had no recognition signals at immediate notice.'[18] Egerton was correct. *V.1* had been allotted a moving haven in which bombing was restricted during the submarine's journey from Lerwick to the Kola Inlet. The position of the 'haven' was adjusted each day to correspond to the submarine's track. To allow for navigational errors a strip 20 miles either side of the haven was made a total bombing restriction area by night with submerged bombing restriction by day. The most likely hypothesis for *V.1*'s end was that she was way out of position, had no recognition signals to hand when the Liberator was sighted and so dived and was attacked. It is worth noting that the aircraft itself was out of position by 90 miles.

A fourth reason is lack of communication among various naval forces operating in the same area. This can lead to confusion and to friendly submarines being identified as hostile. The Battle of Heligoland Bight on 28 August 1914 was a confused engagement with British battle-cruisers, cruisers, destroyers and submarines all milling about in poor visibility in a relatively small area. In particular, the British submarines had not been warned of the presence of Vice Adm Beatty's Battle-cruiser Fleet whose unexpected intervention 'was most embarrassing and caused one or two of the submarines much anxiety'.[19] One submarine to be caused 'much anxiety' was *E.6* (Lt Cdr Cecil Talbot) which 'on two occasions got very close to one of our light cruisers and only refrained from firing when he actually distinguished the red St George's cross in her ensign; until then it did not occur to him that she might be a friend. This shows how close she was and the probability of a successful attack had *E-6* fired. On the second occasion *E.6* was manoeuvred to allow the light cruiser to pass on a parallel course within 300yd in order to make certain of identifying her ensign. Lt Cdr Talbot could hardly be blamed had he attacked, because in this particular operation he was clearly given to understand that only certain vessels were to take part. I submit, however, that his care, patience and good judgement were worthy of praise.'[20]

Talbot certainly acted with considerable restraint. Had he been spotted by one of the cruisers his submarine would probably have been rammed without hesitation. The incident illustrates many of the aspects of friendly fire which will become depressingly familiar throughout this chapter: poor information about disposition of friendly forces coupled with problems of recognition. Coordination between different staffs is always difficult in a pressurised war environment but especially so when different nationalities are involved. A case in point concerns the loss of the Italian submarine *Gugliemotti* in the Mediterranean in 1917. A new submarine, she was proceeding from La Spezia to La Maddalena in Sardinia for trials. While *en route* she crossed the path of a southbound convoy escorted by the British sloop *Cyclamen* who had been warned about U-boats in the area. When the darkened shape of the submarine was seen crossing the sloop's bows from port to starboard, she was immediately identified as a U-boat and rammed. It was only when the survivors were rescued that the mistake was realised. *Cyclamen* promptly made the signal which has achieved notoriety as an example of the black humour which so often attends these incidents: 'Have rammed and sunk enemy submarine. Survivors appear to speak Italian.'[21]

The Italians were not amused, especially as fourteen of the submarine's crew had perished. The Italians had informed the French that the submarine would be crossing the route used by convoys but they found that, despite the fact that the convoy had called at Porto Vecchio in Corsica, the French authorities had not warned those at La Maddalena that the convoy was heading south. The waters around Britain were another area were the ships of three navies (Britain, France and the United States) were operating. The establishment of an American destroyer base at Queenstown in Ireland brought American forces into the same waters in which British submarines were operating. Both were engaged to the same purpose – the destruction of U-boats – but it was inevitable that there would be errors in identification. On 29 February 1918 the British submarine *L.2* (Lt Cdr B. Acworth RN) was sighted by the American destroyers *Paulding*, *Davis* and *Trippe*. Despite giving the recognition signal *L.2* was taken for a U-boat and fired on. Acworth dived the submarine only to receive a severe depth-charge attack which forced *L.2* to a depth of over 300ft.[22] before coming back to the surface. Once again she was fired on and a 3in shell went through the after end of the conning tower (luckily without exploding) before Acworth's signalman could convince his assailants of his identity.

It is particularly difficult to coordinate the movements of submarines and aircraft operating in the same area. In the chaos following the scattering of convoy PQ.17 in the Arctic in July 1942 German U-boats and the Luftwaffe were engaged for some time in sinking the merchant ships. The tactical situation was very confused and changing by the hour. *U334* (KL Wilmar Simeon) had just sunk the freighter *Earlston* when his submarine was bombed by a Ju.88. Sufficient damage was done to cause Simeon to break off patrol and return to Kirkenes under the protection of *U456*.

Perhaps the worst such incident concerns the loss of the *Seawolf* in October 1944. USS *Seawolf*, a Seadragon-class submarine built in 1939, was one of the most distinguished American boats serving in the Pacific. In fourteen patrols and under two commanding officers *Seawolf* had sunk 71,609 tons of Japanese shipping. Her fifteenth patrol, under the command of Lt Cdr A.L. Bontier USN, was a special operation delivering supplies and personnel to the east coast of Samar in the Philippines. Her operation coincided with a Seventh Fleet strike on the island of Morotai and in order to protect American submarines in the area, a submarine sanctuary had been established to the north of Morotai. On 2 October 1944 Bontier reported to Commander Seventh Fleet that bad weather had set him back by a day but that he was within the sanctuary where he was sighted by *Narwhal* on 3 October.

However, on the same day the destroyer USS *Shelton* was torpedoed by the Japanese submarine *Ro-41* commanded by Lt Cdr M. Shiizuka IJN while screening a Seventh Fleet carrier task group. Good damage control

The American submarine Seawolf, *sunk by the destroyer USS* Rowell *in October 1944. (US National Archives)*

saved the *Shelton* from sinking but she could not proceed with the task group, so the USS *Rowell* was ordered to stand by her to render assistance. Meanwhile, the task group despatched anti-submarine patrols and at 11.30 hrs two TBM Avengers were launched from the USS *Midway*. One of these aircraft sighted and bombed a submarine, marking the spot with a dye marker. The attack actually took place within the sanctuary but the aircrew had not been briefed about this, or about the presence of friendly submarines in the area, before taking off. The attack was signalled to the *Rowell* which headed for the spot at top speed. On reaching the area she commenced an Asdic search and gained contact at 13.10 hrs. The destroyer then delivered five attacks using her Hedgehog ahead-throwing-weapon and one using depth-charges.

After the first Hedgehog attack, garbled transmissions were heard on SST but these were disregarded because *Rowell*'s commanding officer considered that the submarine was trying to 'jam' his own Asdic. Another Hedgehog attack was delivered and four or five explosions were observed. Unlike depth-charges, Hedgehog projectiles were fitted with a contact fuse: the bomb had to strike something before it would explode. Debris was blown to the surface including something which resembled a periscope.

Seawolf was never heard from again following her sighting by *Narwhal* and as submarine command at Pearl Harbor began to fit the evidence together, the grim truth emerged. Four American submarines, including *Seawolf*, were in the area when *Shelton* was torpedoed. However, the position given by *Seawolf* in her signal of 2 October reporting herself twenty-four hours behind schedule was only 32 miles from the spot where

Cdr Al Bontier of the USS Seawolf.
(US National Archives)

Shelton was torpedoed. US forces in the area were not told of *Seawolf's* presence, nor were the aircrew told of a submarine sanctuary then in force in which the bombing of submarines was prohibited. The inquiry into her loss concluded that *Seawolf* had been sunk either by the Avengers from *Midway* or by the *Rowell*, although they favoured the latter. The SST transmissions heard by *Rowell* were later identified as the submarine trying to establish contact with the destroyer. The atmosphere in *Seawolf's* sound room as the operators frantically tried to make contact via the SST as the destroyer bored in for another attack can only be imagined.

The inquiry noted that a number of individuals were guilty of errors of judgement but recommended that no disciplinary action be taken. *Rowell's* commanding officer was 'subject to censure' but was not proceeded against after his errors were considered 'due to over-zealousness to destroy an enemy'.[23] The sinking of the *Seawolf* was a clear case of the left hand not knowing what the right was up to. The Seventh Fleet task group had disregarded all the provisions of the sanctuary. *Seawolf*'s signal with her amended position, although sent to the Commander of the Seventh Fleet, was not passed on to the task group operating in the area. To add to the confusion, a Japanese submarine had just proved her presence in her area by torpedoing the *Shelton*. If the *Rowell* and the carrier task group had known that an American submarine was within 35 miles of the position where *Shelton* had been torpedoed, they would have proceeded with greater caution. At that time promulgation of Submarine Position Reports was not required by specific orders from higher authority: it was left to individual task group commanders to decide how much information to pass on about friendly submarines. After the loss of the *Seawolf*, their promulgation became mandatory.

Even in the relative safety of an exercise area a submarine can still be vulnerable. On the morning of 9 March 1945 the destroyer HMCS *Haida* (Cdr R. Welland RCN) and two other destroyers were returning to Scapa Flow from an anti-shipping strike off Norway when they passed close to the battleship HMS *Rodney* and her panoply of escorting cruisers and destroyers. The battleship was conducting AA practice and the three strike destroyers were ordered to join her screen. *Haida* took station on *Rodney*'s starboard beam. Visibility was only about 3 miles in heavy rain squalls so the battleship was not visible to *Haida* all the time but kept looming in and out of the mist. The only ship in clear sight was the destroyer HMS *Zealous* which was 1.5 miles on *Haida*'s port bow. *Rodney* could be heard firing through the mist and shortly after 10.00 hrs, a look-out on *Haida*'s bridge collapsed – shot through the neck by a stray 20mm Oerlikon round (the bullet rattled around the bridge and was retrieved by a signalman). That was the first incident of the day. Cdr Welland later recalled:

This event had barely subsided when our Chief Yeoman, Mackie, quietly but firmly pointed out to me a submarine periscope off our starboard quarter. It was about 800yd off and quickly disappeared. I immediately turned hard to port, went full ahead and ordered a ten charge pattern set to 100ft to be ready.

I assumed that the submarine was making a run at *Rodney* and the cruisers. The Chief Yeoman ordered his men to sound the attacking blasts on the siren, to flash the alarm to whoever was in sight and ordered the radio-room to alert the fleet on voice radio. All of which was done in quick time. I steered the ship on to what I judged to be a good intercept course. The Asdic crew did not gain contact – the ship's speed was approaching 28 knots in a full heeled-over turn since we had all boilers on the line in our earlier hurry to get back into Scapa. I ordered firing and when the pattern was clear of the ship, turned hard to starboard and reduced to 12 knots in an attempt to get an Asdic contact for an accurate follow-up attack. Chief Mackie reported to me that the fleet had executed an emergency turn to port, and that the two nearest destroyers were joining for the hunt.

The submarine surfaced in the middle of the depth-charge boil. My gunnery officer asked permission to open fire – the forward 4.7in mountings were on to the submarine and I was appalled that they would open fire. I had instantly recognised the bow of the submarine as belonging to the British T-class, and had my thumb bent far back as I pushed the 'cease fire' gong. I was very familiar with the T-class – we often exercised together; besides I was a specialist ASW officer, and was supposed to know these things. *Haida* did not fire a shot. The submarine was now lying stopped, 400yd directly ahead, broadside on and badly down by the stern.[24]

By this stage the rest of the fleet had been well and truly stirred up. *Rodney* opened fire with her 6in armament and at least two of the cruisers followed suit. (It was subsequently found that over 150 shells of 6in calibre or greater were fired by the cruisers and *Rodney*.) *Haida* was broadcasting 'cease fire' on every frequency and by any other medium but in the end Welland had to place his ship between the fleet and the submarine. At this stage a Polish destroyer ordered him to clear the range at once or be fired on! Eventually the gunfire subsided (even from the Poles) but it had been a very anxious few minutes.

Welland took *Haida* alongside the submarine and was pleasantly surprised, and not a little embarrassed, to find that it was HMS *Trusty* commanded by Lt P.H. May RN, an old friend of Welland's. 'Tell the bastards to stop firing!' was May's greeting. *Haida* remained alongside *Trusty* for about an hour while the submarine made hasty repairs to the damage, before escorting her back to Scapa Flow where they secured alongside each other. That evening both ships' companies 'spliced the mainbrace' while there was a particularly lively wardroom party.

An informal inquiry was held on board *Rodney* the next day. It transpired that the Duty Staff Officer in *Rodney* had given the order for *Haida* and her consorts to join the screen without informing anyone else on the staff. As a result details of the exercise were not passed to *Haida* and since she had been away on operations she had not received the routine signal about the exercise which was broadcast on the administrative net two days beforehand. As a result none of the four 'strike' destroyers was aware that *Trusty* was in the area. To complicate matters, a U-boat had been reported in the area, which is why the fleet was zigzagging. Welland continues:

This known German threat enabled the trigger-fingered gunners in the cruisers and in *Rodney*'s secondary fire control tower to absolve themselves neatly from any responsibility in shelling the surfaced submarine. The gunners made the points that they were retiring from the scene, the visibility was poor, the range was close to 8,000yd and they couldn't see the target clearly anyway. If the submarine was threat sufficient to warrant a destroyer's depth-charging, what was so wrong in shelling it? They had a point.

Commodore 'D', who conducted the inquiry, publicly absolved *Haida* from any fault. He also congratulated us on the speed and accuracy of the attack, our timely warnings to the fleet, and our blocking the range that stopped the gunfire. Not so fortunate were two of the Admiral's staff. They were reprimanded.[25]

According to *Trusty*'s version of events, she had been observing the exercise from a distance and had watched *Haida* go right by without being detected (the submarine was end-on and thus presented a small Asdic target). May heard the increase in *Haida*'s speed and guessed that he had been seen and so dived to 50ft to allow *Haida* to pass over before firing the recognition signal. The arrival of the depth-charge pattern came as a complete surprise. May later believed that the only thing which saved the boat was that the charges had been set for 100ft. Had they been set any shallower, the submarine would have been sunk. As it was, *Trusty* lost control and had to surface with a good deal of minor shock damage.

The incident was glossed over by both Welland and May, not out of any conspiracy but because it was administratively more convenient to do so:

> This incident was overlooked in official reporting both in the UK and in Canada. Why? Because Commodore 'D' asked May and I (at dinner on the day of the inquiry) if our reporting rules demanded a report. May said that none of his crew had been badly hurt; most of the damage could be fixed in Scapa and that he didn't much like writing reports anyway. I said that my immediate boss was the Commodore himself and that while I sent a monthly report to Canada, the omission of the *Trusty* incident would hardly be noticed.[26]

A final reason for friendly fire incidents is self-preservation by other ships, aircraft and submarines in the same area. If a surface ship detected a dived submarine and there was any doubt about the submarine's identity, it was best to depth-charge and attack first rather than risk being attacked and sunk. During the First World War warship COs and Masters of merchant vessels were encouraged to attack all submarines without hesitation. The results could be unfortunate.

At the end of February 1918 HMS *H.5* (Lt A.W. Forbes RN) departed from Bantry Bay for a patrol in the southern Irish Sea. Nothing more was heard from the submarine and her crew of twenty-five. At 20.30 hrs on 2 March the steamer *Rutherglen* sighted a submarine in position 53° 4'N, 4° 40'W crossing her bows at considerable speed. *Rutherglen*'s Master identified the submarine as a U-boat and rammed her. Cries were heard and men were seen swimming in the water but none was picked up. Since no U-boats were operating in the area at the time, *Rutherglen*'s victim was almost certainly *H.5*. One of those killed was Ensign E.F. Childs of the United States Navy, an American submariner who was on board *H.5* for some war experience. He was one of the first American casualties of the First World War. It is not clear whether *Rutherglen* had been warned that a British submarine was operational along her route. Certainly neither party made any attempt to identify themselves. The Admiralty was not especially concerned about the incident, and *Rutherglen*'s crew were told that they had sunk a U-boat and they received the usual reward. The deception went so far as to sanction the award of the Distinguished Service Cross to *Rutherglen*'s Master, *pour encourager les autres*. The

USS Tullibee *was sunk by one of her own torpedoes which malfunctioned while attacking a Japanese convoy in March 1944. (US National Archives)*

reason for this was that the Admiralty constantly exhorted merchant ship Masters to attack U-boats at every opportunity: the news that a British submarine had been rammed might well make others cautious.

The ultimate form of submarine friendly fire is to be hit by one's own torpedo. Stories of torpedoes which have 'run rogue' after firing are legendary. However, during the Second World War two, and possibly three, American submarines were sunk by their own torpedoes which failed to run true. The USS *Tullibee* (Cdr Charles F. Brindupke USN) left Pearl Harbor on 5 March and after refuelling at Midway departed for her patrol area on 14 March. Her orders were to head to the north of Palau where she was to participate in Operation Desecrate, an attack by carrier-borne aircraft on Palau planned for 30 March. She was not heard from again and was formally declared lost on 15 May 1944. There the matter rested until the end of the war when an emaciated Gunners Mate 2nd Class C.W. Kuykendall was liberated from the Ashio copper mines in Japan. Kuykendall was the only survivor from *Tullibee*'s crew of eighty officers and men, and naturally he was subjected to a thorough debriefing from staff of the Submarine Force as to what had happened to his submarine. He related how on the night of 26 March, he was on look-out duty on the bridge when radar contact was established with a convoy consisting of a troopship and three freighters escorted by a destroyer and two A/S vessels. Visibility was poor with frequent rain squalls. Twice Brindupke began an attack, both times having to break off on account of the driving rain. On his third attempt he drove through the escort screen, by now aware of his presence and dropping random depth-charges, and fired two torpedoes at the troopship. Kuykendall was peering through his binoculars waiting for the explosion when the submarine was rocked by a tremendous explosion and he was thrown into the sea.

Kapitänleutnant Peter Cremer.
(UBA 1411/1/3)

As he struggled to remain afloat he could hear the shouts and cries of other survivors in the darkness around him. He stayed afloat throughout the night and was picked up the next morning by a Japanese escort. Kuykendall was certain that the Japanese had not sunk the *Tullibee*. The ranges and bearings of the three escorts put them out of position for a successful counter-attack. The only other conclusion was that one of the two Mk.XIVs fired at the transport suffered a failure of its steering gear and circled back to strike the *Tullibee*.

A second US submarine, the USS *Tang* (Lt Cdr Richard H. O'Kane USN), a Balao-class fleet submarine with a displacement of 1512/1875 tons, was lost on 25 October 1944 in the Formosa Straits. *Tang* had left Midway on 27 September 1944 for her fifth war patrol. O'Kane was an extremely experienced officer who had commanded *Tang* since her commissioning in 1944 and who had learned his trade as executive officer to the famous 'Mush' Morton of USS *Wahoo*. Under O'Kane's command *Tang* had sunk seventeen Japanese merchant ships totalling 116,454 tons in four patrols. *Tang*'s fifth patrol was set to be as successful as the others. O'Kane chose for his patrol the area between the north-west coast of Formosa and the China coast. This was a focal point for Japanese shipping but was bounded by minefields to the east and the enemy coast to the west. On the night of 10/11 October O'Kane sank the *Joshu Go* and *Oita Maru*, and on 23 October sank a further three vessels (*Toun Maru*, *Wakatake Maru* and *Tatsuju Maru*) in a fiercely fought convoy action.

Twenty-four hours later contact was established with another convoy steaming south to support the Japanese forces engaged in the Philippines. The decks of the tankers and freighters were piled high with crated aircraft. O'Kane decided to attack on the surface and drove through the screen, launching two torpedoes at a tanker and four at two transports. Fired at a range of less than 1,000yd, the six torpedoes hit as planned with a series of violent explosions. *Tang*'s position was revealed by the light of the exploding ships and she was virtually boxed in by merchant ships and escorts converging from all sides. A large troopship and a tanker were astern of her, a destroyer was coming at her from one side while two Kaibokan-type escorts came in from the other. Three burning ships were ahead of the submarine.

O'Kane ordered full speed to clear the area while at the same firing on the tanker, the transport and the closing destroyer.[27] Once clear of the convoy, the torpedomen loaded the last two Mk.XVIIIs into the forward tubes. At 01.25 hrs on the morning of 25 October O'Kane took *Tang* back to finish off the troopship. The escorts were patrolling to seaward of the ship and allowed *Tang* to slip in from the landward side. At 02.30 hrs the range was down to just under 1,000yd and the first torpedo was fired. The second torpedo, *Tang*'s last, was fired shortly afterwards but instead of running true, broke surface, made a sharp turn to the left and began to porpoise off *Tang*'s port bow while turning in toward the submarine.

O'Kane now attempted to take his submarine outside the errant torpedo's turning circle. The helm was put over to starboard, followed by a turn to port in an attempt to swing the stern clear of the warhead. But it was to no avail. The torpedo bore in relentlessly and struck *Tang* on the port side abreast the after torpedo room and close to the manoeuvring-room bulkhead:

> Our stern went under before those of us topside could recover from the blast. One glance aft told me that there would be insufficient time to clear the bridge ahead of the sea. My order 'Close the hatch' was automatic, and my heart went out to those below and to the young men topside who must now face the sea.
>
> Our ship sank by the stern in seconds, the way a pendulum might swing down in a viscous liquid. The seas rolled in from aft, washing us from the bridge and shears, and of small consolation now was the detonation of our 23rd torpedo as it hit home in the transport.[28]

Of the nine officers and men on the bridge at the time the torpedo struck, three, including O'Kane, swam throughout the night and were picked up eight hours later by a Japanese patrol boat. A fourth officer managed to escape from the flooded conning tower and stayed afloat by ingeniously converting his trousers into a life-jacket. Inside the submarine, the explosion of the torpedo was very violent, breaking HP air lines and lifting deck plates. Many of the crew sustained fractures from being violently thrown to the deck. The after torpedo-room, manoeuvring-room and after engine-room were flooded immediately and none of the crew escaped from these compartments. It had been impossible to shut the upper hatch on the bridge and the lower hatch was leaking badly, so the remaining personnel, thirty officers and men, made their way forward to the forward torpedo room, carrying the wounded in blankets, to prepare for an escape.

Despite repeated depth-charge attacks from Japanese escorts, the escape began at 06.00 hrs, nearly four hours after the submarine had sunk. However, only thirteen men actually escaped from the submarine. It is believed that the remainder were asphyxiated by fumes from an electrical fire and the burning of confidential books or a battery explosion. Of the thirteen who escaped, five clung to *Tang*'s indicator buoy and were picked up, three others reached the surface but subsequently drowned and the remaining five were not seen after leaving the escape trunk.[29] Seventy-eight officers and men of the *Tang* were lost.

A third US submarine may also have been the victim of one of her own torpedoes. The USS *Growler* (Cdr Thomas B. Oakley Jr USN) was sunk on 8 November 1944 off Mindoro in the Philippines in position 13° 53'N, 119° 26'E. There are two possible explanations to account for her loss, but insufficient evidence exists to permit a definite conclusion to be drawn. *Growler* had left Fremantle in Western Australia on 20 October 1944 for her eleventh war patrol. On 8 November she was attacked by the Japanese destroyer *Shigure* and escort vessels *Chiburi* and *No. 19*. The Japanese made their attacks in their usual desultory manner; there was no visible result and no claim was made for a submarine sinking. Distant 'observers' to *Growler*'s ordeal were the submarines *Hake* and *Hardhead* which were occupying neighbouring patrol positions. Both heard the depth-charging on their passive sonars and *Hardhead* also heard the detonation of a torpedo.

At least two U-boats are known to have fallen victim to their own T.5s. In January 1944 *U377* (Oblt.z.S Gerhard Kluth) and *U382* (Oblt.z.S Rudolf Zorn) were engaged in operations north of the Azores where the American carrier group TG.21.16 was operating in support of slow GUS (Gibraltar–USA) and UC (UK–Caribbean) convoys. *U377* is often claimed as having been sunk by escorts of TG.21.16, but these attacks were on *U382* which survived. At 04.04 hrs on 15 January *U377* reported having unsuccessfully attacked TG.21.16 with a T.5. Shortly after that signal a further message, with a corrupt signature, was received by U-boat Command to the effect that the originator had been hit by a torpedo, was badly damaged and was sinking. The signal was sent on the Diana frequency, one of three frequencies used by U-boats when engaged in operations, and therefore the signal had to come from a U-boat. *U377* was never heard from again, which was unusual given that since the boat was in contact with a task group it would have been reasonable to expect further situation reports from her. Since there is no evidence to link any other boat with the signal, it is most likely that the originator was *U377*.

The second such U-boat to be sunk by a T.5 was *U869*. The circumstances of her loss are unclear but the recent discovery of her wreck with a gaping hole by the submarine's control-room, the area from where the most noise would be emanating when the boat was dived, would seem to indicate that a rogue T.5 was responsible.

NOTES

1 Capt John Coote to author, 17 May 1989.
2 Cdr Arthur Pitt to author, 12 May 1989.
3 The submarine did not have to surface fully to use the 267 radar. It was possible to use the set while submerged by running with only the aerial, which was mounted on a fixed mast at the forward end of the periscope standards (the brackets supporting the periscopes above the conning tower), above the water.
4 Vice Adm Sir John Roxburgh to author, 13 March 1989.
5 PRO ADM199/721. Reports of the commanding officers of ORP *Jastrazb*, HNorMS *St Albans* and HMS *Seagull*; pp. 99 *et seq*.
6 DNC's *Submarine War Experiences*, Report of HMS *Talisman*, p. 109.
7 Capt John Stevens to author, 10 February 1989.
8 Stevens, ibid.
9 PRO ADM137/3745: Commodore British Adriatic Force (CCBAF) to Admiralty, 25 April 1918.
10 Simpson, Rear Adm G.W., *Periscope View, A Professional Autobiography* (London, Peter Davis, 1965).
11 PRO ADM137/3745: Report of the Board of Inquiry into the loss of *H.5*.
12 PRO ADM137/3745: CCBAF to Admiralty, 25 April 1918.
13 RNSM *Thrasher File*: Ken Sunor to Jan Lock, 26 June 1989.
14 Vice Adm Sir Hugh Mackenzie to author, 7 February 1989.
15 DNC's *Submarine War Experiences*, Report of HMS *Thrasher*, pp. 106–8. Typescript copy in RNSM library.
16 RNSM *Thrasher File*: Ken Sunor to Jan Lock, 26 June 1989.
17 Cloarec's account of the last moments of *La Perle* is preserved in the French naval archives in Paris: file no. SHM TTY.771.
18 PRO ADM.199/1104, SBNO *North Russia Report of Proceedings No. 3*, June and July 1944.
19 Keyes MSS 4/30: Keyes to Rear Adm Christian, 29 August 1914.
20 Ibid.
21 Halpern, Paul, *The Naval War in the Mediterranean 1914–1918* (London, Allen & Unwin, 1987), p. 339.
22 In 1925 the officially sanctioned maximum diving depth for an L-class submarine was 150ft – the depth to which they were tested was 100ft.
23 Roscoe, Theodore, *US Submarine Operations in the Second World War* (USNIP, 1988), p. 418.
24 Rear Adm Robert Welland's account dated 15 May 1986, RNSM Archives.
25 Ibid.
26 Ibid.
27 O'Kane, Rear Adm Richard, *Clear the Bridge! The War Patrols of the* USS *Tang* (New York, Rand McNally, 1977), p. 453.
28 O'Kane, op. cit., p. 456.
29 For the record, the escape of the men from USS *Tang* was done from a depth greater than any had ever escaped from before using the Momsen Lung escape apparatus.

ASW – AWFULLY SLOW WARFARE?

ASW is the last of the warfare areas that has not lent itself well to automation, such as the Aegis combat system. You still have to think – generally ahead of your opponent – and all the time; it is perhaps the last great chess game.

Vice Adm J.R. Fitzgerald USN

The development of anti-submarine warfare techniques was an inevitable follow-on to the development of the submarine. Submarines existed, they posed a threat to warships and merchant shipping alike, and therefore means had to be found of dealing with them. In one sense this cycle was no different from any other in the naval race that preceded the First World War – weapons had been devised, such as the torpedo boat, and then means had been found to counter them. However, the submarine possessed one advantage which made it doubly hard to find and attack: once submerged it was invisible. In the words of Sir Henry Newbolt, the official British historian of the war at sea, 'by the end of March 1917 there had been one hundred and forty two actions between German U-boats and British destroyers and the destroyers had only sunk their opponents in six of them. When therefore a German submarine commander fell in with a British destroyer, though perhaps he would have to submerge and chance his ground, still his chances of escaping destruction were about 23 to 1.'[1]

Of all the chapters of this book, this is the one that is most clearly delineated on national grounds. In the twentieth century the story of ASW has largely been written by the Royal Navy and US Navy effort against German U-boats in both world wars, and against the threat of Russian submarines in the postwar period. In both world wars German/Italian/Japanese ASW was not accorded a high priority – with one exception.

The story of ASW in this century has run along two parallel tracks. The first is the design and development of sensors to detect and track a submerged submarine. The second is the development of weapons/tactics to deal with a submerged submarine.

It was in the use of sound to detect a submerged submarine that the most progress was made. There are two approaches to using sound under water: passive and active. Passive requires the use of an underwater microphone to detect and classify the sounds made by a submerged submarine. In its crudest form this meant dangling a microphone, known as a hydrophone, over the side of the hunting ship, which lay stopped during this procedure so that her own hull and engine noises did not deafen the hydrophone's operator. Active required the transmission of a beam of sound through the water. Certain materials, known as transducers, change shape when an electric current is passed through them. This results in a sharp burst of sound being produced – the 'ping' that is such a feature of fictional submarine films. If the 'ping' strikes a submerged submarine (or a wreck or other underwater object) it is reflected back towards its point of origin. The speed of sound being known, a calculation based on the exact times of transmission and reception of the echo produces an indication of the submarine's range and position. Transducers can also be used in the passive role. When they are struck by a sound wave, they produce a weak electric current which can be used to power a display.

Active measures, or Asdic,[2] were never deployed at sea during the First World War but were developed extensively in the interwar period. In particular, the Royal Navy saw Asdic as the universal panacea for dealing with the submarine threat in another war. Early British sets produced a narrow searchlight beam. The operator pointed his transducer in one direction, sent out a 'ping' and then listened for an echo before searching in another direction. Maximum effective range was just over a mile, although this could be affected by water conditions. It follows that such a set was more useful for attack than detection since the set could be used for rapid searching in many directions at once. However, in August 1937 an incident occurred which showed the limitations of Asdic and suggested that perhaps the British were placing excessive faith in its capabilities. During the Spanish Civil War the Italian submarine *Iride*, operating under Spanish control, attacked the British destroyer *Havock*. *Havock* attempted to attack her assailant using her Asdic set to gain contact. Despite a calm sea and good water conditions, contact could not be held and *Iride* got away.

During the Second World War the limitations of Asdic were rapidly exposed. The stylised exercises with tame submarines off Portland were shown to have little relevance in the North Atlantic in winter. To begin with, U-boats operated on the surface at night, thus making Asdic nearly redundant. Otto Kretschmer, Germany's leading U-boat Ace, was only sunk in March 1941 because an inexperienced OOW dived the boat instead of escaping on the surface, and thus exposed *U99* to HMS *Walker*'s Asdic.

The development of Asdic during the Second World War was accorded a high priority but it is a story which lies beyond the scope of this chapter.[3] Depth-finding sets solved the problem of having to guess a submarine's depth, and thus made weapon delivery a more exact science. However, the introduction of the fast, deep-diving nuclear submarine rendered all Asdic development to date redundant. The development of ASW systems to deal with these new submarines was therefore of major importance. The postwar period would see the development of medium- and long-range sonars,[4] variable-depth sonars (where a transducer is lowered from a ship, submarine or helicopter to beneath a temperature layer), convergence-range sonars and passive sonobuoy and towed-array systems. Two further

HMS Hesperus *arrives in the Gladstone Dock in Liverpool in January 1943 with damage to her bows sustained when she rammed U357 on 26 December 1942. Ramming was a satisfying but ultimately pyrrhic way of sinking a submarine.* Hesperus *would be unavailable for operations for several months while the damage was repaired. (Author)*

growth areas in sonar were broad-area sensors such as SOSUS laid across the GIUK gap, a choke-point for Soviet submarines entering the Atlantic.

Ramming was the first tried-and-tested method of anti-submarine warfare. The cruiser HMS *Birmingham* showed the way when she caught *U15* on the surface off Fair Isle in the North Sea on 9 August 1914 and rammed her. This was a very satisfactory way of sinking a submarine. The result was highly visible and there was lots of excitement for all concerned. However, it could produce a somewhat pyrrhic victory. An escort with damaged bows would be unavailable for operations for months while the damage was repaired – which was perhaps not unwelcome to the ship's company concerned but hardly favoured shipyards already overloaded with wartime work. In one instance ramming proved fatal for both submarine and attacker.

On 31 May 1918 *UC51* (ObLtz.S W. Schmitz) attacked an east coast convoy but was damaged by the freighter *Blaydonian* which ran over her casing. As the U-boat came to the surface she was attacked by the escort, the destroyer HMS *Fairy*. *Fairy* rammed the U-boat twice, the first time at the stern and a second time between the conning tower and the deck gun. However, *Fairy*'s bows were badly damaged in the two rammings. She had been launched in 1897 and had just rammed and sunk an opponent one-and-a-half times her size and considerably more stoutly built. Damage control could not stem the flooding and she sank shortly afterwards. In the Second World War the Royal Navy advised against ramming, but they could hardly prohibit a practice which accorded so well with the Nelsonian diktat that, 'no captain can do very wrong who lays his ship alongside that of the enemy'.

How was this problem to be dealt with? The matter of locating submarines underwater would not be solved for some time. Much time and resources were spent on methods which, viewed with the benefit of hindsight, were hopelessly unrealistic. Early in the First World War one early ASW 'expert' remembered this delightful scheme which was carried out by patrols in Portland: 'The anti-submarine armament consisted of a blacksmith's hammer and a canvas bag. The idea was that the hammer should be used to smash the periscope of the submarine, and the canvas bag to blind the submarine by being tied over the top of the periscope.'[5]

Others advocated the use of specially trained seals to detect and indicate the presence of a submerged submarine – an option not unknown in our time! Another scheme advocated covering the North Sea in foaming bath salts which would force submarines to the surface where they could be dealt with.[6] Other early ASW measures owed more to cavalry tactics than to naval warfare. Patrols carried out at high speed by destroyers in areas known to be frequented by hostile submarines were popular, as was the deployment of massive numbers of auxiliary warships in aimless and undirected searches. In restricted waters, such as the Black Sea, Adriatic and Baltic, these tactics might have worked. But in the wide oceans areas of the Atlantic, they were a waste of effort. The reason for this failure is not hard to find. The hunters are relatively large and can be seen from a great way off, while a submarine on the surface presents a very small target.

Some of the first ASW measures concentrated on self-protection rather than destruction of the enemy. Before and during the First World War all capital ships were fitted with anti-torpedo nets while at anchor. These ungainly contraptions, suspended from booms around the ships, had to be raised and stowed – a time-consuming and tedious process – before the ship got under way. Nets were also used in static boom defences at harbour entrances.

Geography played into the hands of the British during the First World War. The use of the term British is not meant to disparage Britain's French and Italian allies, however it must be admitted that it was Britain which provided most of the drive and nearly all of the *matériel* in terms of ASW effort. Britain or her allies controlled the vital choke-points through which U-boats would have to pass on their way to the high seas: the Dover Straits, the Scotland–Norway gap, the Straits of Gibraltar and the Straits of Otranto at the southern end of the Adriatic. Blocking these routes with minefields might keep the U-boats penned in their harbours. Tremendous effort went into devising and laying minefields and net barrages in these areas. The blockade across the Straits of Otranto was never really effective, and no wonder for it was an ambitious undertaking: the Straits are over 40 miles wide, the depth of water is over 3,000ft in the middle and the area is subject to sudden storms of great intensity.

The Otranto barrage originally consisted of drifters towing indicator nets. As it became clear that U-boats were still getting through, the barrage forces were repeatedly strengthened until in February 1918 over 200 ships and 70 aircraft were involved in its maintenance. Moreover, the barrage constituted a tempting target for Austrian light forces based at Cattaro and there were a number of tip and run raids. By March 1918 the barrage consisted of a submarine patrol off Cattaro, a destroyer screen further south, then two lines of trawlers between Cape Linguetta and Corfu Island equipped with hydrophones and depth-charges, and finally, a line of American submarine chasers working from Corfu. Despite this impressive array, U-boat commanders still boasted that they could slip through on the surface at night. In over three years of war only two submarines were sunk by the barrage forces – the Austrian *U6* on 13 May 1916 and the German *UB53* on 3 August 1918. Two other submarines, the German *UB44* and Austrian *U30*, disappeared and may have fallen victim to mines while outward-bound through the Straits. A fifth U-boat, *UB52*, was sunk by the British submarine *H.4* to the north of the barrage on 23 May 1918.

The Dover Straits barrage was perhaps the most important because it tried to control the movements of U-boats from the Belgian bases of Ostende and Zeebrugge. Up till the end of 1917 the barrage was barely worth the effort expended on it. It consisted of minefields supplemented by indicator nets towed by trawlers and backed up by destroyers. By the end of 1917 it was recognised that the barrage was utterly ineffective (documents retrieved from sunken U-boats gave explicit instructions on how to get through it!), and it was replaced by a deep minefield running across the Dover Straits from Folkestone to Boulogne. This minefield was to be intensively patrolled by day and night. U-boats attempting to pass through would either be forced to dive into the minefield and destruction, or battle it out on the surface. The new arrangements paid immediate dividends. Not only were a greater number of U-boats sunk, but they were now forced to make the longer journey 'north-about' around the top of Scotland into the Atlantic rather than going the 'quick way' down the English Channel. This in itself represented a victory since U-boats were spending more time and fuel on passage rather than in operational areas.

The blocking of the Dover Straits was followed by the most audacious blocking attempt of all: the Northern Barrage. Minefields laid from the Orkney Islands to the limits of Norwegian territorial waters (Norway being neutral) off Hardanger Fjord to the south of Bergen would prevent U-boats using the 'north-about' route. This was an American proposal, and agreement to it at the inter-Allied conference in London in September 1917 was contingent on the US providing the necessary mines (some 100,000) since factories in Britain were fully stretched. It aroused considerable opposition from the British Grand Fleet which claimed that the mines (and the new American Antenna mines were distrusted) would do no more than severely restrict the Grand Fleet's sea room in the event of a German break-out.

The barrage consisted of three areas: Area A, the central area, was 30 miles long and laid with shallow mines; Area B, the western sector, was laid with deep minefields and patrolled so that U-boats would be forced to

*Laying mines in the English
Channel during the First World
War. Mines accounted for 58 of the
178 U-boats sunk during the First
World War.*

dive into the mines; and Area C, nearest the Norwegian coast, was laid
with both deep and shallow mines. In September 1917 the Norwegian
government, under British pressure, declared that they would extend Area
C throughout their own waters using British-supplied mines so that
U-boats could not take a short cut through Norwegian waters. Minelaying
began in Area B on 3 March 1918 and in Areas A and C on 8 June. The
barrage consumed 70,263 mines and cost $40 million – at 1918 prices. It
claimed six U-boats sunk and possibly a seventh with two or three
damaged. The northern barrage was typically American: tremendous
effort, a grand design and hugely expensive. But it has to be said that the
results were certainly not in proportion to the effort.

The first proper ASW weapon was the depth-charge. This was a
derivative of the explosive sweep – an explosive charge which was towed
behind a warship and could be fired when it encountered an underwater
object. Such a weapon was responsible for the destruction of the Austrian
U23 on 21 February 1918 by the Italian destroyer *Airone*. However, the
sweep was ungainly and proved to be a complicated weapon which could
not be carried by all ships. The depth-charge was a canister of explosive,
fitted with a hydrostatic fuse, which was dropped over the stern of the
attacking ship. Later refinements meant that depth-charges could be

The characteristic 'boil' on the surface left by depth-charges. All too often this was the only mark left over the position of a sunken submarine. (Author)

thrown outboard from the sides of the attacking vessel and dropped by aircraft. The charge then sank to its pre-set depth where it exploded. All submariners, of whatever conflict or nationality, experienced the same fears and emotions during a depth-charge attack. The attacking ship could be heard steadying up on her course and then roaring overhead. There would then follow a brief pause before the charges exploded, shattering gauges, breaking woodwork and throwing men all over the boat. If the submarine was deep and a sea-water connection fractured, it did not just leak: the water streamed into the boat like an iron bar, with sufficient strength to break a man's arm.

This description of a depth-charge attack on 11 June 1917 comes from the IWO of the Austrian submarine *U27* (Linienschiffsleutnant Robert Teufl von Fernland). The submarine had just attacked a flotilla of Japanese destroyers in the Aegean and had blown the bows off the *Sakaki*. Retribution from the Japanese was now about to be administered:

I heard the explosion of our torpedo like discreet hand-clapping only, but this shot of ours opened the gates of hell. In vain we turned away from the wounded destroyer, in vain we went as usual down to twenty five, then fifty, then seventy-five metres as a series of bombs exploded over and around us. First the lights went out, then the gyro compasses stopped working, soon afterwards the packing rings started to leak. In this situation we could not risk using the pumps too frequently, only in the very rare intermissions between attacks. We once went down to eighty-five metres then suddenly up to twenty five again and higher, because the skipper wanted to know what had happened to his target. New series of bombs followed, more water poured into the boat. Down we went again.

After the torpedo attack had been launched, I had nothing to do. I tried to get some rest on my bunk. At times I observed our men. They quietly and silently carried out the skipper's orders, given by a few gestures of his head or hands only. The enlarged shadows of the crew appeared in the emergency light on the wet walls of the boat like enormous ghosts sentenced to perpetual drudgery in a watery inferno. Sometimes we had to risk pumping again because the water did not stop dripping in. We had so much water in the boat that it really was not easy to keep the boat at a depth that the walls of the submarine would stand. Hours of anxiety and watchfulness passed.

U27's ordeal lasted 11½ hours. She survived but many were not so lucky.

The first U-boat to be sunk by a depth-charge was *U29* on 13 December 1916. The most depth-charges expended in a single hunt

The end of U744 on 6 March 1944 after a hunt which had lasted for more than twenty-four hours and in which over 291 depth-charges had been expended. U744 was ultimately sunk by a torpedo fired by HMS Icarus. (Public Archives of Canada)

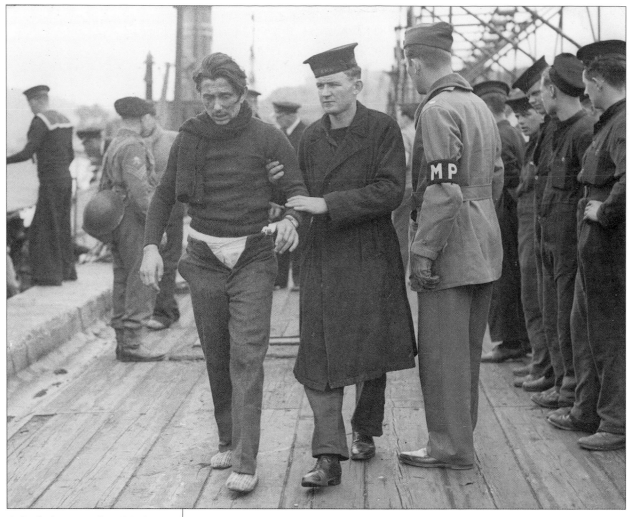

A shocked survivor from U269, sunk on 25 June 1944 by HMS Bickerton in the English Channel, is landed at Southampton. (Author)

against a U-boat is believed to be over 290 dropped on *U744* on 6 March 1944.

Submariners' reactions to depth-charge attacks varied. The more phlegmatic accepted them as the inevitable consequence of an attack. There was little that the majority of the crew could do other than relax (to conserve oxygen) and await the inevitable. Panic or moral collapse was unheard of in all navies. A few men manifested some odd signs of behaviour indicating strain – in HMS *Upholder* one man began a loud and fervent recitation of the Lord's Prayer in the control-room as the charges went off around him. Generally most men waited and sat the attack out, aided by a little courage out of a bottle. British submariners hoarded their rum ration in 'blitz bottles' for precisely such an occasion, while in American submarines little 2oz bottles of whiskey – issued for strictly medicinal purposes – had a habit of circulating during an attack. The violence of a depth-charge attack could also confuse men about the true state of damage to their boat. After *U570* (KL Hans Rahmlow) had been straddled by four depth-charges dropped by a Hudson of 269 Sqn on 27 August 1941, Rahmlow was so confused that he thought his boat was more damaged than in fact she was. This appreciation led him to surrender.

The commanding officer, of course, was the one man who could not simply sit out the attack. He was the focus of attention in the boat and it was on his skill at evading the attacker that the boat's survival often depended. The experienced CO soon learned to listen for the sound of the attacking ship steadying on her course and then to make radical alterations of depth and course. British and American submariners argued about the merits of remaining (relatively) shallow or deep while under attack. Some argued that at shallow depth the boat had more residual strength left in the pressure hull to absorb the punishment. Others, particularly German U-boat commanders, argued that safety lay in the depths. A Type VIIC U-boat could dive to 200m, well below most depth-charge settings. This fact was appreciated early in the war and the more ingenious TGMs began to block the hydrostatic fuses of their depth-charges with soap to allow them to sink to beyond the pre-set depths before exploding. Eventually the British developed the Mk.XX depth-charge, a one-ton monster that had to be fired out of a torpedo tube, and which sank to great depths before exploding.

Problems with the delivery of the depth-charge, in particular the loss of contact just before the weapons were dropped, resulted in the development of ahead-throwing-weapons like Hedgehog and Squid. Hedgehog was a 24-spigot anti-submarine mortar firing 7in diameter contact-fused projectiles weighing 65lb with a 35lb charge. It was

The British submarine J.1 – the only submarine to drop a depth-charge on another submarine (possibly UB51), on 9 November 1918. This remains a unique feat in submarine history. (Author)

designed to be fired ahead of the attacking ship and thus gave the advantage that the ship did not lose Asdic contact with the target during the final attack run. It was fitted to destroyers and frigates and by the end of the Second World War had been fitted to 500 ships. Squid was a more advanced weapon. It consisted of a three-barrelled 12in mortar with the barrels fixed one behind the other in a frame that could be tilted through 90° to bring the barrels horizontal for loading. The projectiles weighed 390lb with a Minol charge weighing 207lb. Sinking speed was 43.5f/s and clockwork time fuses were used, set automatically from the depth recorder. Training was possible to 30° either side of the bow but the weapon was meant to be fired dead ahead. It was automatically fired from the Asdic range recorder and threw the three bombs to land in a triangle with 40yd sides 275yd ahead of the ship. The successful installation of Squid depended on the ship having a 'Q' attachment and the later Type 147 Asdic which gave an estimation of the target's depth. The first successful sinking by Squid was that of *U736* by HMS *Loch Killin* on 6 August 1944. KK Peter Cremer, who commanded *U333*, said that 'Hedgehog was a hand probing for the U-boat with five fingers; the Squid was a paw which struck out and crushed everything'. In the postwar period the weapon underwent a variety of modifications and emerged as the Mk.X mortar widely fitted in British and Commonwealth ships.

The introduction of aircraft into the ASW armoury was a major step forward. It was the Austrians who made the first significant use of aircraft in the ASW role in the clear blue waters off the Dalmatian coast. And it is

Lieutenant de Vaisseau *Henri Devin* (centre left), commanding officer of the French submarine Foucault, seated on the Austrian seaplane L.135 *with his captors. (Kriegsarchiv, Vienna.)*

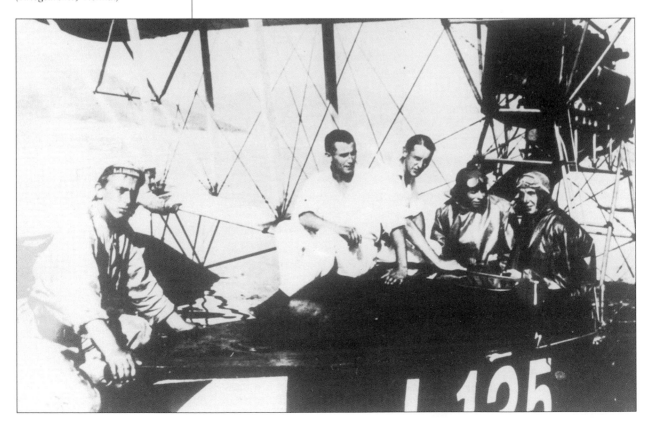

The French submarine Foucault *which was sunk on 15 September 1916 by Austrian aircraft was the first submarine to be sunk in a deliberate air attack. (Marius Bar)*

to them that honours go for the first deliberate sinking of a submarine by an aircraft.[7] On 15 September 1916 the French submarine *Foucault* (LV Devin) was spotted while dived off the port of Cattaro by an Austrian seaplane. The Austrian aircrew were unsure of the identity of the submarine and did not attack – a rare occurrence for aircrew! Instead they returned to Cattaro and, after confirming that no friendly boats were in the area, returned to the scene, this time armed with two 50kg bombs. After searching for half an hour the submarine was located and Fregattenleutnant Walter Zezelny dropped both bombs alongside the submarine's port quarter. On board *Foucault*, the first indication that anything was amiss came when the bombs exploded. The motors were damaged and the stern glands began to leak, causing a short circuit and an electrical fire. Devin reluctantly ordered his boat to the surface where she was abandoned. Zezelny landed alongside and took Devin and his first lieutenant back to Cattaro as a 'sample'. The remainder of *Foucault*'s crew were picked up by a torpedo boat.

Early aircraft were slow, ungainly and poorly equipped to attack a submarine. However, an aircraft – or airship or tethered kite balloon – possessed the advantage of height and was able to spot a surfaced submarine long before the submarine could move into position for an attack. The mere presence of an aircraft was often sufficient to deter a submarine and this is an important feature of ASW often neglected by many commentators. Diesel-electric submarines (in pre-schnorkel days) depended on being able to surface and charge their batteries in peace. If they were forced to dive and 'break the charge', then their operational capability would be reduced considerably.

Aircraft were a major ingredient in ASW during the Second World War. Land-based VLR aircraft like the Liberator could range far out into the Atlantic. Working in conjunction with a convoy escort, such an aircraft,

Happy days in the Mediterranean in 1917 before aircraft made their presence felt as an ASW force to be reckoned with.
Linienschiffsleutnant *Georg von Trapp of the Austrian* U14 *dictates his* Kriegstagesbuch *to his IWO. (Kriegsarchiv, Vienna)*

especially one fitted with ASV radar, could locate U-boats far away from the convoy, attack them and then guide in an escort ship for the kill. Shipborne aircraft could operate in areas beyond the reach of land-based aircraft and could respond immediately to a threat.

Aircraft required specialised lightweight ASW munitions. Shipborne depth-charges were too heavy and the small A/S bombs with which British aircraft were equipped at the beginning of the Second World War tended to break up on impact.[8] The most significant development in terms of airborne ASW weaponry was the acoustic homing torpedo. Known, predictably, as 'Fido', for security reasons the device was known coyly as

The dangers of air attack. The British C.25 under attack by German seaplanes on 6 July 1918 off Harwich. C.25's pressure hull was holed, she was unable to dive and she was only saved by the appearance of another British submarine which drove her assailants off. (Author)

L98, an Austrian Lohner seaplane. The Austrian Navy was the first to use aircraft extensively for ASW, considerably aided by the crystal clear waters on their side of the Adriatic. (Kriegsarchiv, Vienna)

the Mk.XXIV mine. Fido's guidance was based on four passive hydrophones. When dropped into the water from an aircraft Fido sank to a pre-set depth and began an acoustic search for the submarine, the detection range being 1,500yd. If nothing was 'heard', the torpedo commenced a circular search pattern which was maintained for up to fifteen minutes. The weapon was first used by US Navy TBM Avenger torpedo-bombers operating from escort carriers in the Atlantic and its first success was the sinking of *U266* on 14 May 1943 by a VLR Liberator of 86 Sqn RAF.

Some observers were more than a little caustic about Fido's performance. However, when discussing the sinking of *U525* (KL Hans-Joachim Drewitz) on 11 August 1943 Captain Arnold J. Isbell USN of the escort carrier USS *Card* commented: 'It is believed that the Mk.24 Mine is far more effective then commonly given credit. It is acknowledged that our eggs are in one basket. However, when the mine functions as it is supposed to – the show is over and there is no further argument by either the enemy or ourselves.'

Other weapons developed for airborne use included rockets and powerful cannon such as the 57mm Tsetse gun fitted in some RAF Beaufighter and Mosquito aircraft. The variety of weapons available to aircraft enabled them to develop specialised tactics for dealing with submarines and the Americans were particularly adept at this. Aircraft working in pairs would carry different weapon loads. F4F aircraft would carry cannon and rockets for flak suppression and to force the submarine to dive. The second aircraft, usually a TBM Avenger, would then either drop depth-charges as the submarine was diving or, once the submarine had dived, drop a Fido. The introduction of sonobuoys gave aircraft such as the Avenger an autonomous hunting capability. Faced with such tactics, the cards were well and truly stacked against the submarine. The commander had either to remain on the surface and fight it out or dive and risk attack by a Fido. In more modern times the commander of the Argentine *Santa Fe* (CC Hector Bicain) faced precisely such a dilemma on April 1982 when his submarine was attacked off South Georgia by British helicopters. On this occasion Bicain elected to fight it out on the surface – he knew that if he dived, one of the helicopters would be bound to drop a homing torpedo – and eventually staggered into Grytviken with his fin shredded by British missile hits and machine-gun fire.

Along with the tremendous development of individual weapons came changes of tactics. The most significant development was that of the creeping attack, where one escort held the Asdic contact while directing another ship to make the attack. This meant that the U-boat could not make evasive moves at the last moment. However, if a U-boat chose to remain on the surface, then it would be immune to Asdic detection. Great effort then went into the development of radar for escorts. Once detected by radar, a U-boat could be illuminated by flares, and then attacked with gunfire and forced to dive. Once a U-boat had dived she was effectively removed from the convoy battle, a 'mission kill': she could not observe events and report and lacked the underwater speed to keep up with the action.

As British and, particularly, American shipbuilding got into their stride, greater numbers of anti-submarine vessels became available. Hence it was

A low-level photograph showing U848 during attacks by four US Navy PB4Ys and two USAAF B-25s on 5 November 1944.
U848 endured nearly a day of such attacks before blowing up at 16.00 hrs. Twenty survivors were seen in the water but only one was
rescued after spending nearly a month in a liferaft. (US Navy)

Lt.z.S H. Schauffel, the 2WO of U752 and the only commissioned survivor, on his arrival in England. U752 was sunk by rocket (the first U-boat to be sunk by rockets) and cannon fire by aircraft from HMS Archer on 23 May 1943. Schauffel had a lucky escape – both the CO and the IWO were killed standing beside him. (US Navy)

possible to form support groups, a group of escorts not attached to a particular convoy but which could be used to protect a number of convoys by responding to intelligence indications about U-boat activity. The Americans went one better and formed 'hunter-killer' groups based around an escort carrier. These groups did excellent work in the central Atlantic. In the case of the support and the hunter killer groups, training was the key to performance so that each commanding officer and his team knew their place in the overall picture. Thus the development of work-up facilities at shore bases where escort vessel crews could be rigorously drilled in shore simulators was of major importance.

Attacking the enemy submarines in their bases was another area of ASW. Often seen more as a part of the strategic bomber offensive, sinking a U-boat alongside at Hamburg was just as good as sinking one in the North Atlantic.

One notable ASW measure was the convoy. Trade protection measures, notably the introduction of the convoy system, achieved concentration of ASW assets (in an escort group) and forced the U-boats to come in search of their prey. In doing so they exposed themselves to detection, attack and possible destruction. Lastly, the intelligence war, in the form of breaking

enemy codes and the development of HF/DF equipment, allowed patterns of enemy submarines to be monitored and individual U-boats, unwise enough to be detected transmitting while on the surface, to be prosecuted.

Thus ASW in the Second World War was not just a simple ship vs submarine affair. It was the synthesis of a vast number of different skills and applications, involving more than one service. The resources devoted by Britain and America toward ASW permitted them to engage in their choice of strategy on land. The fact that the U-boats had been controlled not only allowed the landings in Normandy in June 1944 to take place, but permitted the Allies to land at a place and time of their choosing. On the other side of the world, the Japanese neglected ASW and thus American submarines achieved the annihilation of the Japanese merchant marine, cutting Japan off from her conquests and more importantly from her sources of food and raw materials.

It must be stated, however, that the sum of the ASW effort during the Second World War merely succeeded in driving the submarine from the surface. And while '1945-style ASW' was well placed to deal with the submersible torpedo-carrier, it was not tried against the submarines which the Germans were about to deploy at the end of the Second World War – the Type XXI and Type XXIII which possessed enhanced qualities

This remarkable photograph taken from the island of the escort carrier USS Guadalcanal *shows flying operations in progress while the carrier tows the German U505 captured on 4 June 1944. Two days later when Eisenhower's communiqué concerning the Normandy landings was posted on the ship a sailor commented, 'Boy! Look what Eisenhower had to do to top us!' U505 is now permanently displayed in Chicago. (US Navy)*

A US Navy escort carrier in mid-Atlantic in 1944. Such ships were able to deploy air power out to areas beyond the reach of shore-based aircraft. (US Navy)

A remarkable photograph showing the end of a Geman U-boat. The 'dots' on the forward face of the conning tower are the aerials for a fixed radar array – to train the set the boat had to be turned. (US Navy)

A Type VII U-boat under air attack during the last days of the Second World War. (US Navy)

of underwater speed and endurance. In the sole reported encounter between a Type XXI U-boat and an Allied escort group (after the German capitulation) the U-boat made a dummy attack and then escaped with consummate ease.

In the postwar period ASW has been dominated by two factors. First, the growth of a large Soviet fleet of submarines – 320 submarines of the Whiskey, Zulu and Foxtrot classes were completed between 1950 and 1970. This fleet was targeted at the Atlantic supply line, the means by which America would reinforce and supply her armies in Europe in the event of war. The large fleet of SSKs was followed by an equally capable fleet of nuclear submarines armed with cruise missiles for strikes on western carrier groups and with ballistic missiles for attacks against Nato members' home territories. This placed ASW at the very top of NATO's agenda, so much that after Britain's withdrawal from east of Suez in the 1960s and before the collapse of the USSR in 1990, the Royal Navy was busily transforming itself into a specialist ASW force for the eastern Atlantic. The second factor was the development of the nuclear-powered submarine. The potential of the SSN or SSBN demonstrated the futility of nearly all that had been achieved in ASW since the First World War.

The chief problem was that the development of sonar had reached such a stage that detection was now theoretically possible at great distances from the attacking ship. However, weapons such as Squid and Hedgehog were restricted to the immediate area around the ship. What was wanted was some form of carrier to take the weapon out to the desired range. One option was the nuclear depth-bomb, a weapon that would destroy anything within a large radius of the point of detonation. However, a more productive approach was to develop weapons with a greater range and some form of mid-course guidance after the weapon entered the water. ASROC (Anti-Submarine Rocket) was an American ASW missile which was designed to deliver a homing torpedo as close as possible to the target submarine in the minimum time, thus limiting the submarine's time to take avoiding action or launch a counter-attack. It consists of a homing torpedo, a Mk.44 or Mk.46, which is fired from a launcher on the deck of a ship to a range programmed by the ship's sonars. After a predetermined flight time, the motor shuts off and the torpedo drops into the sea and activates. A nuclear version was also developed using a 1KT warhead. A version for use by submarines was also developed called SUBROC (Submarine Rocket) which was first deployed in 1965. The missile was fired, broke surface and then flew to a point (maximum range was 28.5 miles) above the submerged enemy submarine as indicated on the firing submarine's sonar. It then dropped a 1.5KT W55 nuclear depth-charge. It was withdrawn in 1988.

The Drone Anti-Submarine Helicopter (DASH) was an imaginative but unsuccessful American attempt to employ an unmanned helicopter to

The British frigate HMS Brazen. *This ship is fitted with towed array passive sonar together with a sophisticated active sonar suite. She also carries a Lynx helicopter armed with homing torpedoes and constitutes a deadly ASW unit. (Royal Navy)*

deliver a homing torpedo to within 200yd of a submerged submarine 10,000yd away. Many ships in the USN and JMSDF[9] were fitted to carry DASH but the system was prematurely withdrawn, since it proved extremely difficult to control the drones in flight. Of the 746 DASH drones built, over half were lost at sea in accidents.

The final solution to this problem lay with the introduction of a helicopter small enough to be carried on the flight deck of a frigate. The use of helicopters has been one of the most significant ASW developments since the war and is perhaps the one most feared by the submariner. They are fast, carry a lethal payload and from the submarine's point of view, are undetectable. The helicopter was conceived as little more than a carrier to take the weapon out of range of the ship's sonar. However, helicopters such as the British Wessex Mk.3 and Soviet *Hormone* had their own dipping sonars which they could use to locate a target. In the future, helicopters such as the Anglo-Italian EH-101 will be capable of autonomous search, location and attack at considerable distances from the parent ship. One British submarine developed an interesting way of dealing with an ASW helicopter. During an exercise in the 1950s HMS *Springer* (Lt P.R. Compton-Hall RN) was being chased by a helicopter with a dipping sonar. Compton-Hall simply rammed the transponder, damaging it beyond repair.

In the postwar period the submarine has come into its own as the ASW platform par excellence. Britain had developed specialised 'submarine-hunters' during the First World War – the R-class with their high underwater speed of 15 knots. However, these specialised submarines had not survived the tremendous run-down of the Royal Navy which followed

A British R-class submarine built in 1918 and one of a class of twelve submarines built specifically to hunt and destroy other submarines. (Wright & Logan)

the Treaty of Versailles. During the Second World War there had been a number of submarine vs submarine engagements, nearly always with one party on the surface.[10] However, the development of first the schnorkel and then nuclear power meant that submarines were no longer bound by the ritual of the daily surface to charge the battery.

The modern SSN is the most potent ASW platform in existence. Her nuclear powerplant gives her unlimited power with which to run sophisticated active and passive sonars which can track targets many hundreds of miles away while she lurks quietly in position. Homing torpedoes, either thermally or electrically powered, can be used to deliver the attack. Modern Russian submarines now carry a rocket torpedo, travelling at a speed of 200mph plus, which gives the opposing submarine barely any time at all to react before destruction.

NOTES

1 Newbolt, *Naval Operations*, vol. 1, p. 342.

2 There is some confusion over the origin of this acronym. Most authorities, including this author, held that it stood for *Allied Submarine Detection Investigation Committee* but no trace of the existence of such a committee can be found. It is most likely a corruption of *Anti-Submarine Division-ics*.

3 The interested reader should consult Hackmann, Willem, *Seek and Strike* (London, HMSO, 1989).

4 Sonar – US acronym for Sound Navigation and Ranging. This is now common throughout English-speaking navies and has replaced Asdic.

5 Edwards, Lt Cdr Kenneth, *We Dive at Dawn* (London, Rich & Cowan, 1939), p. 369.

6 This scheme is not as far-fetched as might be thought. As a defence against midget submarine attack on submarine bases, some have advocated covering the harbour with fuel oil. The environmentalists would howl but it would be an effective counter to periscope observation.

7 The British submarine *B.10* had been sunk in an air raid on Venice on 9 August 1916 but the attack had not been deliberate – the Austrians were bombing the dockyard and *B.10* got in the way!

8 One RAF aircraft attacked a British S/M in error with these bombs. The bombs exploded on impact with the water, fragments of the bomb flew upwards and did such damage to the 'bomber' that the aircraft was forced to ditch.

9 Japanese Maritime Self Defence Force: the coy term for what is now the third largest navy in the world. None of its ships is more than ten years old and they are armed with the most sophisticated weapons and sensors that money and/or the Japanese electronics industry can provide.

10 The exception being the sinking of *U864* on 9 February 1945 by HMS *Venturer*.

SPECIAL
OPERATIONS

The submarine's ability to conceal itself makes it an ideal recruit to the secret world. Submarines have participated in any number of special operations of many varieties – and still continue to do so. The range of these operations is seemingly endless and includes landing and retrieval of special forces/agents, operating with midget submarines, gun-running, beach reconnaissance, acting as navigational beacons and surveillance.

Landing and recovering personnel from an enemy coast is a tricky operation. Expert navigation is required to bring the boat to the right position. If there is a 'reception party' waiting on shore, there must be pre-arranged identification signals to ensure no unpleasant surprises. During the First World War there was little call for this sort of operation. The most celebrated submarine 'passenger' was the Irish Nationalist Sir Roger Casement who was landed by the Germans in southern Ireland with two companions. Casement was returning to Ireland having failed to win support from the Germans for the cause of Irish Independence. He sailed on 12 April 1916 from Wilhelmshaven in *U19* (ObLtz.S Weisbach) and was to be put ashore in Tralee Bay, where Weisbach was to rendezvous with the coaster *Aud* which was carrying 20,000 rifles and over 1,000,000 rounds of small arms ammunition for Sinn Fein. The signal that all was safe was to be two green lights burned from the shore or a pilot vessel.

The operation was nothing less than a catalogue of disasters. Both the *Aud* and *U19* were in Tralee Bay on the evening of 20 April, yet Weisbach made no attempt to contact the *Aud* even though she was clearly visible. Worse, there was no signal from the shore indicating all was well. The Sinn Feiners were not expecting either vessel for another forty-eight hours and the two green lamps were still hanging unlit in the Drill Hall at Tralee. Eventually the *Aud* moved away and headed for Lisbon and safety.[1] *U19* waited for some time, hoping to see the signal, before Weisbach decided that with dawn coming he could not remain in the area for much longer. He gave Casement the choice: either to land and take his chances ashore or to return to Germany with the submarine. Casement chose the former and was put ashore with his companions in a rubber dinghy.[2] The operation illustrated many of the pitfalls of such an undertaking.

During the Second World War the landing and recovery of agents formed a significantly greater proportion of submarine operations than during the First World War. Following the conquest of large areas of

Europe and Asia by the Axis powers, Britain and America formed the Special Operations Executive (SOE) and Office of Strategic Services (OSS) respectively, which were specifically charged with encouraging and organising resistance in these areas. In the words of SOE's charter the organisation was responsible for 'All action by way of subversion and sabotage, against the enemy overseas.'[3] The problem lay in getting the agents, couriers, special forces or supplies to the operational areas which were cut off from Britain, or other Allied command centres, by sea. The air route was one option but was very vulnerable to interception and could not carry supplies in bulk. The sea route was the alternative and seemed to offer numerous advantages. SOE maintained its own 'private navy' of disguised fishing smacks and motor gunboats for precisely this purpose. However, submarines were the ideal craft. They could carry stores in bulk and were discreet: a submarine could make the 'delivery' and then retire into the depths.

The case for using submarines for these operations was very finely balanced. From one perspective, using a submarine for such purposes was ideal. Yet from the naval perspective there were equally compelling arguments why submarines should not be used. Each submarine was an important unit of the fleet with a crew of about sixty officers and men. Submarines were important assets which could not be risked lightly in support of what many naval officers (of all navies) regarded as 'boys' own adventures' which appeared to have little bearing on the war effort. A particular bone of contention in Anglo-American circles was that while engaged on a special operation, the submarine was prohibited from engaging legitimate enemy targets for fear of compromising the operation.

On 2 May 1944 Lt T.G. Ridgeway, commanding HMS *Templar*, on patrol in the Malacca Straits, was mortified at having to allow a 3,000-ton unescorted merchant ship – a rare target in those waters – to pass unhindered as he was engaged in transporting an SOE reconnaissance party. 'Another case of a target lost because of a special operation,'[4] wrote *Templar*'s flotilla commander, Capt H.M.C. Ionides RN. The apparent inability of the various intelligence organisations to cooperate was another difficult area. Thus HMS *Tally-Ho!* (Lt Cdr L.W.A. Bennington RN) found herself embarking parties from two different organisations, SOE and the ISLD (Inter Services Liaison Department – the section of MI6 responsible for operations in the Far East). Disagreements between the two parties on board resulted in the SOE officer vetoing the landing of the ISLD party even though Bennington considered that the landing was safe.

Thirdly, there was the inevitable falling-off of efficiency which accompanied a submarine's deployment on special operations. Commenting on the decision to use three T-class submarines to carry 'chariots' to attack targets at Palermo and Maddalena in January 1943, the Naval Staff History concludes: 'These Chariot operations cost the loss, directly or indirectly, of two T-class submarines, both with experienced crews, and ten 'charioteers', which must be weighed against the destruction of one light cruiser and damage to an 8,500-ton liner. Furthermore the diversion of submarines for Chariot carrying and recovery duties gravely interrupted their normal patrol activities at a time when there were many valuable targets at sea on the Axis supply routes to North Africa.'[5]

A Chariot surfaces with its two-man crew. (Author)

Writing in equal vein following the use of three U-class submarines as Chariot carriers in connection with beach reconnaissance of Sicily in May–June 1943, Capt G.C. Phillips RN, captain (S)10 noted: 'It is to be hoped that these reconnaissances are of real value to the planners. It has become clear from recent results that submarines engaged on reconnaissance work fall off in efficiency for offensive patrols.'[6]

Lastly, there was always the fear that the landing site would be compromised and the submarine sunk or captured. A submarine lying on the surface in shallow water off an enemy coast with hatches open for the despatch/reception of the party was very vulnerable. In May 1944 HMS *Storm* (Lt E. Young RNVR) had to land and recover an agent off the island of Puloh Weh, north of Sumatra. The landing went without a hitch

In May 1944, HMS Storm *sailed into a Japanese ambush after recovering an agent from an island north of Sumatra.*

but the recovery four days later was a near disaster. The agent had been captured and *Storm* had sailed into a Japanese ambush. She came under heavy machine-gun and artillery fire from the shore and only managed to extricate herself by diving in very shallow water. Young commented afterwards, 'It was remarkable that we had got away without a single casualty, but the most surprising feature of the whole incident was that we had seen no enemy ship of any kind. Perhaps the Jap soldiery were jealous of their naval forces and hoped to win the kudos of sinking us unaided. If so they had been very stupid: properly organised land and sea cooperation should have had us absolutely cold.'[7]

These risks aside, submarines of all the major belligerents were employed to varying degrees in these operations. The submarines of the Royal Navy were the most extensively thus employed, landing/recovering agents in France, Norway, Italy, Greece, Crete, North Africa and the Far East. American submarines were also extensively used for this purpose in the Pacific. Other Allied nations such as the Poles, Norwegians, French and Dutch also used submarines to land/recover agents for their own purposes as well as the overall Allied cause. The Germans, too, used U-boats to infiltrate the occasional agent into the United States. There is no record of Italian, Soviet or Japanese submarines participating in this kind of operation.

A different but related kind of operation to those described above is the use of submarines to land organised military forces ashore. The first examples of these operations involved the submarine sending members of her crew ashore to destroy a particular objective. In August 1915, during operations in the Sea of Marmara, Lt Cdr Nasmith of HMS *E.11* sent ashore his first lieutenant, Lt D'Oyly Hughes, to destroy a railway bridge at Ismid using demolition charges. After hearing a loud explosion *E.11*'s crew were relieved to see their first lieutenant swimming out to them.

Apart from the dangers inherent in any such operation, there are no passengers on a submarine and to risk losing a vital member of the ship's company for a comparatively unimportant target was perhaps unwise. In the same month the submarine *E.2* (Lt Cdr de B. Stocks RN) attempted a similar operation. Lt L. Lyon, the first lieutenant, was sent ashore to blow up the Constantinople–Adrianople railway where it ran close to the sea west of the capital. Although a loud explosion was heard, nothing was ever seen or heard of Lyon again, and nothing could be learned from the Turks after the war. There were no further such operations.

As in so many aspects of submarine development it was the Italians who took the lead in this particular field. The stalemate in the Adriatic during the First World War provided the impetus for a number of daring and innovative schemes for attacking the Austrian Navy in their bases. As part of a series of plans for attacking the main Austrian base at Pola, the submarine *Argo* was converted to carry frogmen, whose task was to dismantle the net barrages at the entrance to Pola harbour to allow either MAs (Motoscafi Antisommergibile – fast torpedo-carrying motor boats) or other submarines to enter the harbour. For this purpose *Argo*'s fore ends were converted to a single 'exit–re-entry' compartment allowing the frogmen to leave and re-enter the craft while the vessel remained submerged. The First World War ended before the operation could be carried out.

During the Second World War the Italians deployed the élite Gamma frogmen who operated from submarines (among other places), using limpet mines to sink Allied shipping. The Regia Marina developed quite sophisticated techniques for these operations. In an attack on Algiers in December 1942 the submarine *Ambra* (CC Mario Arillo) carried ten Gamma swimmers along with two Maiale (the Italian SLC two-man human torpedo). In an effort to improve target selection TV Augusto Jacobacci would leave the submarine via the forward escape hatch and swim to the surface carrying a telephone connected to the submarine by a strong armoured cable. Using the telephone Jacobacci was able to guide Arillo to the best location for launching the Maiale Midget submarine. The Gamma men then left the submarine and swam to the targets indicated to them by Jacobacci. It was also hoped that Jacobacci could help in the recovery of the frogmen but this was not to be – all ten were captured, but four merchant ships were damaged and the Allies were forced to implement tighter port security.

The British experience was confined to submarine cooperation with organised military forces such as the newly formed Royal Marine Commandos and their related organisation, the Special Boat Service. Perhaps the best-known such operation is an attack on shipping in Bordeaux by twelve Royal Marine Commandos under the command of

An Italian Maiale *being raised from the water in 1942. Note the shrouded propeller which was intended to ensure trouble-free passage through nets and other obstructions. (Dott. Achille Rastelli)*

Maj 'Blondie' Haslar: Operation Frankton. The marines were equipped with folding two-man canoes known as 'Folboats' and a large number of explosive charges. They were conveyed to their dropping-off position off the mouth of the Gironde Estuary by HMS *Tuna* (Lt R.P. Raikes RN) and launched on the evening of 7 December 1942. The raid, immortalised in the film *Cockleshell Heroes*, was a success with four merchant ships damaged.

US submarines supported raiding operations in the Pacific, the best-known example being the attack on Makin Island on 17 August 1942 by the Second Raider Battalion transported there by USS *Nautilus* (Lt Cdr W.H. Brockman USN) and *Argonaut* (Lt Cdr J. Pierce USN). The raid was a success – but only just. *Nautilus* had to supply some unplanned fire support from her two 6in guns when the soldiers became bogged down ashore. The operation illustrated a number of disadvantages in using submarines for this purpose. Both submarines were continually bothered by aircraft during the day and eventually had to submerge, leaving the soldiers ashore until nightfall when they could be recovered. The submarines were woefully unprepared to deal with the wounded – an inevitable consequence of land operations – and conditions in *Nautilus'* wardroom, which doubled as the operating theatre, resembled a scene from the Crimea.

In May 1943 *Nautilus* took 109 Army Scouts to Attu on the Aleutians in advance of the main US landings. The scouts all suffered from sea sickness and by the end of the day the air in the boat was so foul that even though oxygen was released into the boat, there was not enough in the atmosphere to light a cigarette. By the end of trip the scouts would have gladly taken on the entire Imperial Japanese Army rather than remain on board for another day!

This operation illustrates some of the problems associated with submarines working with 'commando-style' forces. Submarines by their nature are fairly small and space for the carriage of a military party and their equipment is limited. It is one thing for a submarine to insert a single agent or pair of agents. It is quite another thing to carry twenty or thirty commandos, their small arms, supplies and support weapons. Yet if a raiding party is to pack any kind of punch it must be fairly large and have sufficient organic firepower to get out of trouble. Submarines are also unable to stay and rescue the party if they get into trouble – *Nautilus'* ability to supply 65 rounds of 6in shells was a fortunate coincidence.

Lastly, the submarine has a role to play in the transport of military supplies to foment rebellion in occupied territories or among ethnic minorities. During the First World War German U-boats maintained regular supply runs taking arms and ammunition to the Senussi tribe in North Africa. A religious sect, the Senussi had been waging war against Britain's ally, Italy, since 1912 when Italy had wrested Cyrenaica from Turkey. Three UC-type submarine minelayers, *UC128*, *UC20* and *UC73*, were employed exclusively on transporting arms to the Senussi. These operations brought the German Navy into sharp conflict with the General Staff who considered that the deployment of these submarines as little more than military transports was more important than unleashing them against Allied merchant shipping. However, the German efforts to supply the Senussi resulted in what must be one of the most unusual submarine passengers. In August 1918 Obltz.S H. Rohne of *UC20* was presented by the Senussi with a goat, a camel and a sheep as gifts for the Kaiser. How the animals were shoehorned into *UC20*'s 52m long hull is not recorded although it must be presumed that the camel was a young one. Conditions inside the small submarine with her 26-man crew and animal 'guests' can but be imagined. On arrival at Pola the camel was found to be too sick for onward transit to Berlin and lived out its days on the island of Brioni. What happened to the goat and sheep has not been recorded although given the shortage of meat in the Dual Monarchy, their fate can be imagined. In 1942 both *U339* and *U274*, engaged on similar operations, received similar 'presents', which both commanding officers got rid of when tactfully out of sight of the donor!

During the Second World War both British and American submarines carried arms in support of various guerrilla groups, particularly in the Far East. The record for such an operation goes to the British submarine *Thule* (Lt Cdr A.C. Mars DSO RN) On 6 February 1945 she landed 4,000lb of equipment for the MPLA (Malayan Peoples' Liberation Army) at Johore, north of Singapore. On 25 May 1945 she was back, this time with 8,000lb of stores for the MPLA. In order to land all this equipment *Thule* embarked 3 officers and 16 Royal Marines and carried 14 rigid raiding craft and 11 outboard motors. With all this equipment on board, there was no room for any reload torpedoes – a situation that Mars disliked

The Israeli submarine Dakar *leaves Portsmouth in November 1967. She subsequently foundered in the Eastern Mediterranean, cause unknown. Note the very large fin which contained a five man exit–re-entry chamber for special forces insertion and/or recovery. Israeli submarines spend a high proportion of their time operating in support of special forces. (Wright & Logan)*

enormously. He was even more disgruntled at the lengthy unloading procedure insisted on by the Marines. Mars merely wanted to unload the munitions and get clear; the Marines wanted the weapons unloaded in a clear sequence. The majority of the weapons were marked solely for the MPLA and had been adapted so that they would 'wear out' after firing a certain number of rounds. Thus they could not be used against the British in any postwar colonial conflict. The remainder of the cargo comprised unaltered weapons for the use of the MPLA's 'advisers'.

In the postwar period there has been no lessening in the type of landing/recovery operations carried out by submarines allied to a continual refinement of techniques. Submariners have always been able to use escape chambers and hatches as an exit from the submarine but the postwar period has seen the construction of dedicated 'exit–re-entry' chambers, which mean that the submarine no longer has to surface to launch the 'operators'. This eliminates the most dangerous part of the operation from the submariner's perspective. The first submarine to have such a fitting was the Israeli *Dakar* (ex-British *Totem*) which was refitted to incorporate a five-man 'exit–re-entry' chamber in her fin. Subsequently,

US Navy Seals release an SDV from a dry dock shelter on the American submarine USS Kamehameha during exercises off Hawaii. Supporting special forces is one of the many functions of the modern submarine. (US Navy)

the British Oberon-class SSKs *Otus* and *Opossum* were refitted to carry similar chambers in their fins.

Ideally, operators lock in or out of an escape hatch or dedicated hatch and use electrically driven SDVs (Swimmer Delivery Vehicles) to get ashore. These vehicles can be stowed in chambers outside the pressure hull in very much the same way as Italian Maiale and British Chariots were carried. The launch can thus take place quite close inshore without the submarine having to surface. If the submarine has to surface to launch the operators and their craft, it may be possible to surface some distance offshore and then tow the craft by using a periscope to engage a wire strung between the two craft.

Such operations have been well practised by the major submarine powers. In 1982 HMS *Onyx* was the only British SSK deployed to the South Atlantic at the time of the Falklands War specifically for the landing/recovery of special forces. In the Gulf War of 1991 HMS *Oracle* was likewise despatched to the Persian Gulf for the same purpose, but the war ended before she could be thus employed. However, these operations are increasingly the preserve of the 'second division' submarine powers.

Pakistan possesses a human torpedo capability and the Indians expected an attack by Pakistani submarines using these weapons during the Indo-Pakistan War of December 1971. In the event, the Pakistani command chose to send their submarines armed with conventional torpedoes. Israel in particular sees the landing/recovery of special forces as a primary operational requirement for its small submarine fleet as did pre-Mandela South Africa. Both countries face (or faced in the case of South Africa) significant internal security problems and see their small submarine fleets as important elements in the counter-insurgency role. (Israel currently has three Type-206 German-designed but British-built SSKs and South Africa has four French Daphne-class SSKs.)

The other side of the coin is that submarines are admirably suited to foment discontent. Fears exist that Iran will use its newly acquired Kilo-class SSKs to stir up religious trouble in Saudi Arabia – only a short distance across the Persian Gulf. North Korea makes considerable use of its submarine fleet in its never-ending war with South Korea. On 17 September 1996 a North Korean Sang-O submarine ran aground near Kangnung in South Korea. Only one of the twenty-three crew members survived – the others were either killed by their own side in an execution-style shooting or in fighting with South Korean forces. The submarine was carrying a team of commandos who were planning to launch an attack on a forthcoming international economic conference in South Korea. When the submarine was examined she was found to contain a 107mm rocket launcher, a 75mm anti-rocket launcher and 190 other weapons including AK47 and M16 assault rifles. The submarine was also found to be equipped with a lock-out chamber.[9]

Beach reconnaissance prior to an amphibious landing is another unusual operation performed by submarines. This can take place either by direct periscope reconnaissance of the beach, or by landing specially trained observers to note beach defences, gradients, soil samples and other details required by the planners. These are generally considered hazardous operations as they require the submarine's prolonged presence off an enemy coast in shallow waters which may well be mined. The first recorded beach reconnaissance was, however, more mundane.

In August 1917 the British were preparing to land an infantry division at Westende on the coast of Belgium with the aim of capturing Ostende and, it was hoped, turning the flank of the German Army. Planning for the operation was hampered by the total lack of any hydrographic data on the landing area. Aerial observation and clandestine trips by survey launches could fill some of the gaps but what the British needed to know was the range of the tidal rise and fall over a number of weeks. Accordingly, the old submarine *C.17* (Lt C. Wardell-Yerburgh), was ordered to proceed across the channel and lie submerged on the bottom off Nieuport, to register the height of water above her hull continuously for twenty-four hours by reading the depth gauge. The rise and fall and the tide curve at this spot was thus obtained at springs, neaps and intermediate tides. It was a weird experience for the submarine to lie submerged off an enemy coast and it was also extremely uncomfortable. *C.17* was not large and was fitted with a petrol engine. Even though she sailed with a reduced complement, some men were nearly unconscious through CO_2 poisoning when she surfaced after a 24-hour period spent submerged. The data obtained was invaluable and won Wardell-Yerburgh a well-deserved DSC, not to mention securing a 'first' for *C.17*.

In the Second World War it fell to the US submarine *Nautilus* (Lt Cdr R.B. Lynch USN) to carry out the first photographic beach reconnaissance operation when in 1943 she photographed three landing sites on the Gilbert Islands. A bracket for holding the camera was attached to the search periscope and the lower sound-room had been converted into a temporary darkroom since immediate processing of the negatives was imperative in case retakes were required. Three cameras were provided but all proved unsuitable largely because vibration of the periscope made photography impossible unless a high shutter-speed was used. This was equally impossible since loss of light through the periscope meant that slow shutter-speeds were essential. Fortunately, Lynch had his own German-made Primaflex camera on board which did the job perfectly.

Other wartime efforts at periscope photography relied on nothing more than placing the lens against the eyepiece of the periscope and hoping for the best. The results were often indifferent and useless for anything other than souvenir value. The incorporation of still cameras into search periscope masts has made this task much easier. During the 1982 Falklands War HMS *Onyx* was able to provide much detailed periscope photography evidence of likely landing sites on the Falklands.

Periscope reconnaissance of a hostile shore can be very risky since it involves the submarine moving at very slow speed in daylight with the periscope raised. Other ways in which submarines can support beach reconnaissance is by acting as support for specialist landing parties such as the British COPP parties or the American UDT or present-day SEALs. In preparation for the invasion of Sicily in the summer of 1943 the British U-class submarines *Safari*, *Unrivalled*, *United*, *Unruffled*, *Unshaken*, *Unseen* and *Unison* all took COPP parties on a total of eighteen visits to various Sicilian beaches. In January 1944 the midget submarine *X.20* took a three-man COPP party for a four-day reconnaissance operation off the Normandy beaches. Each night *X.20* (Lt K. Hudspeth DSO, RANVR) would close the beach and the 'Coppists' would swim ashore to collect their data, including soil samples which they stored in specially de-lubricated condoms. During the day the submarine would lie on the bottom and the crew and passengers would get such rest as they could. After three days of this routine, with five men crammed into the tiny X-craft along with wet clothing and a mass of equipment, Hudspeth had had enough, and came home.

*A British X-craft, in this case X.24,
in Loch Cairnbawn, after her
successful mission to Bergen in
April 1944. (Author)*

One of the strangest tasks ever performed by a submarine was
Operation Mincemeat, a deception operation carried out by the British in
April 1943 in order to confuse the Germans about future Allied operations
in the Mediterranean. The plan, devised by the Naval Intelligence
Department, was to deposit a corpse, dressed in British uniform, on a
Spanish beach. The corpse would be carrying forged documents indicating
the future course of Allied strategy which it was hoped the Spanish would
read and communicate to the Germans. It took some persuading to
convince a dubious Admiralty about the scheme but eventually approval
was given.

A corpse was procured from a London mortuary, the family being
assured that the body was required for 'active service' and that he would,
in time, receive a Christian burial. The body was then given a completely
new identity as Maj William Martin of the Royal Marines, official number
148228, there being a number of William Martins in the Navy List.
'Maj Martin's' pockets were filled with the usual detritus anyone carries
around on a daily basis: money; a photograph of his 'girlfriend' (an
Admiralty typist); theatre ticket stubs and a letter from his bank manager
complaining about his overdraft! However, the *pièce de résistance* was
contained in his briefcase, which was packed with beautifully contrived
correspondence from the Chiefs of Staff to Allied commanders in the
Mediterranean indicating that Sardinia and the Peloponnese rather than

HMS Seraph *with her crew and highly decorated Jolly Roger.* Seraph *participated in a number of special operations including Operation Mincemeat.*

Sicily were the targets for the next Allied invasion. The plan was for 'Maj Martin' to be put in the sea from the submarine *Seraph* (Lt N. Jewell RN) off the Spanish port of Huelva. The position had been chosen in the knowledge that the tide would carry the body ashore. It was hoped that the Spanish would conclude that 'Martin' was the victim of an air accident. It was also hoped that the Spanish would not be able to resist opening the briefcase and passing on the details to the Germans. Complete with his correspondence, 'Maj Martin' was packed in an air-tight canister labelled as meteorological instruments and despatched north to the Clyde and HMS *Seraph*.

On the morning of 30 April 1943 *Seraph* broke surface off Huelva. As soon as Jewell considered it safe, he sent the look-outs below and ordered the body, still in its canister, to be brought up on to the casing. He then sent all ratings below. Jewell and his officers then removed the body and, after checking that the briefcase was securely chained to the corpse's wrist, slid the body into the sea. Jewell deviated but once from his orders, and that was to recite such words as he could remember from the burial service as the body was slid into the sea.

The sea duly deposited 'Maj Martin' ashore at Huelva where the Spanish authorities found the body and reported it to the British naval attaché in Madrid who reported the matter to London on 2 May: 'RO Huelva reports body identified as Maj W. Martin RM, card number 148228, has been washed ashore at Huelva. Death due to drowning probably 8 to 10 days ago. Spanish Naval Authorities have possession of papers found.'[10]

The attaché, Capt Hillgarth, was not privy to the operation so London's response, which arrived on 4 May directly from the DNI (Director of Naval Intelligence and graded 'Immediate',[11] came like a thunderbolt: 'Some of papers Maj Martin had in his possession are of great importance and secrecy. Make formal demand for all papers and notify me by personal signal. . . . Such letters should not, repeat, not be opened or tampered with in any way. If no official letters are recovered make searching but discreet enquiries. . . .'[12]

This signal was followed by another in similar vein forty-eight hours later and Hillgarth would not have needed to be a genius to conclude that London was extremely concerned about the whereabouts of Maj Martin's briefcase. His concern to recover the briefcase intact was instrumental in persuading the Spanish that the documents were real. The Spanish authorities behaved as predicted and granted the Germans a sight of the documents on 9 May – an appreciation of their contents was circulated within OKW on 14 May – before they were returned to the British naval attaché in Madrid. Signals intelligence told the British that the Germans had been totally deceived by the documents. As a result German forces in Italy and Sicily were dispersed and the landings on Sicily later in the year met comparatively little opposition. 'Maj Martin' (and indirectly *Seraph*) had played his part[13] in the Allied victory.

NOTES

1 *Aud* was scuttled the next day after being discovered by the sloops *Bluebell* and *Zinnia*.

2 Casement was arrested less than twenty-four hours later. He was charged with high treason and, following a sensational trial, was hanged in Pentonville Prison.

3 Foot, M.R.D., *Special Operations Executive* (London, BBC Books, 1984), pp. 20–1.

4 Cruikshank, G., *SOE in the Far East* (Oxford, OUP, 1983), p. 144.

5 Admiralty: Naval Staff History, *Submarines. Vol. 2, Operations in the Mediterranean* (London, 1955), p. 120. The submarines concerned were *Traveller* and *P.311*.

6 Op. cit., p. 123.

7 Young, E., *One of Our Submarines* (London, Rupert Hart-Davis, 1953), p. 258.

8 *UC12* made just one voyage as an arms carrier in December 1915 before being converted back to a minelayer. On 16 March 1916 she was sunk off Taranto – blown up on her own mines.

9 Belke, Cdr Thomas J., 'Incident at Kangnung', *Submarine Review* (April 1997).

10 PRO ADM1/25230. NA Madrid to Admiralty, 2 May 1943.

11 The use of the 'Immediate' priority designation, the second highest classification, meant that the signal was of significant importance.

12 PRO WO 106/5921. DNI to NA Madrid, 4 May 1943.

13 'Maj Martin', whose true identity has never been revealed, now rests in the municipal cemetery at Huelva. On his tombstone is the legend, *Dulce et decorum est, pro patria mori.*

BREAKING THE BLOCKADE

The blockade is one of the oldest forms of naval warfare and consists of little more than a campaign to starve the enemy into surrender by cutting off his seaborne trade. It follows that blockade breaking is an equally venerable activity. During the Revolutionary and Napoleonic Wars, Britain effectively cut Napoleon's Europe off from the rest of the world, inspiring the famous quote by the naval historian Mahan that, 'Grass grew on the quays at Antwerp'.

During the First World War Britain moved swiftly to establish a similar blockade on Germany. The threat posed by mines and torpedoes meant that it would not be a close blockade as traditionally practised. Instead the Navy would seal off the various choke-points, such as the Dover Strait, the Greenland–Iceland–Faroes–UK gap and the North Sea, to intercept and detain merchant vessels bound for or leaving Germany. Diplomatic measures, pressure on neutral countries and an ingenious re-definition of just exactly what cargoes constituted contraband[1] all served to tighten the screw on Germany.

And it was extremely effective. Not only was Germany kept short of many strategic materials required for the war effort but she was also cut off from the crucial imports of nitrates without which the country's predominantly sandy soil would become exhausted within five years. Starvation was staring the Germans in the face and conditions worsened as the war went on. Rations became smaller and deliveries of food increasingly irregular. Undoubtedly the effectiveness of the British blockade underpinned the Germans' decision to adopt unrestricted submarine warfare.

The Germans considered every option, practical or otherwise, as to how the blockade could be circumvented. They faced two problems: the first was breaking through the ring of patrols which the British had established at every entrance to the North Sea. The second was that the British also maintained patrols off every major port – even neutral ones like New York. The patrols may have been discreet and out of sight of land but they were still there and constituted a considerable obstacle. Among the wilder schemes was one for cargo ships to discharge their cargoes at sea on to U-boats which could make the rest of the passage to Germany submerged. This notion was swiftly rejected. The use of operational U-boats for cargo-carrying was likewise impractical owing to restrictions on visits by belligerent warships to neutral ports. However, the *Admiralstab* did

The view looking forward down on to the casing of a German UCIII class submarine showing the external torpedo tubes on both sides of the hull. In both world wars the Germans used such spaces for carrying stores. (Author)

consider a proposal from a Bremen businessman, Alfred Lohmann, for the construction of cargo-carrying U-boats.

Lohmann offered to form a limited liability company, together with the Reichsbank and the shipping line Norddeutsche–Lloyd, with a capital of RM.2,000,000. The investors would be guaranteed a return on their money (and 5 per cent interest) by the State, which would then keep any ensuing profits. By November 1915 the terms were agreed and the company established as Deutsche Ozean–Reederei GmbH. Agents of the company were soon established in America and began purchasing cargoes for shipping to Germany. The submarines, which were of projected 1,400-tons surface displacement were the largest submarines built by a German shipyard and presented significant problems. Although the absence of armament simplified the internal arrangement, no one could predict how such a large submarine would handle at sea. At the same time the urgency of the situation meant that no time could be wasted on research and development. Engines and motors from existing submarines would have to be used.

The submarines *Deutschland* and *Bremen* were the result of this venture – with five more planned. At RM 2.750,000 each (£137,000 at 1916 prices), they were not cheap. They had a displacement of 1512/1875 tons, and were 213ft long with a beam of 29ft. Two-shaft 6-cylinder 4-stroke 800bhp diesels gave a speed of 12.4 knots on the surface while two 800shp electric motors gave a dived speed of 5.2 knots. The submarine had a surface range of 25,000nm at 5.5 knots. The cargo of 700 tons was carried both inside the hull and externally. A crew of fifty-six was recruited from operational U-boat men and commercial seamen. For the purposes of the voyage the seamen would possess the usual papers of a merchant seaman. As if to confirm their non-belligerent status, the crew were provided with ample provisions, luxuries not normally seen on operational U-boats, such as cigars and gramophone records, and superb quality clothing. Capt Paul König, an experienced Norddeutsche–Lloyd Master with an extensive knowledge of the American coast, was appointed to command *Deutschland*. None of the crew was under any misapprehensions about the voyage: 'To make money, of course!' was the opinion of Boatswain Humke.

On 23 June 1916 *Deutschland*, flying the German mercantile ensign, sailed from Kiel carrying mail, precious stones and 163 tons of concentrated dye. This last item alone was worth $1,400,000 on the American market. König successfully avoided all the patrols, despite the British having been tipped off by the Dutch merchant ship *Westerdyk* which sighted and reported the submarine on 24 June, and made most of the passage on the surface. However, König was prudent enough to dive whenever smoke or masts were sighted. His caution was justified. Vice Adm Sir George Patey, C-in-C North America and West Indies Station, had announced: 'I cannot conceive that any enemy's submarine can have any status except that of a belligerent enemy and I shall treat her accordingly if met. The fact of flying a merchant flag, or any other flag, would be no guarantee.'[2]

Unfortunately, Patey's ships never 'met' the *Deutschland* which signalled her arrival off the Quarantine Station at Baltimore on 9 July. The agents of the Deutsche Ozean–Reederei done a good job not only in gathering cargo for the return voyage but also in mobilising public opinion. When she finally came alongside at Baltimore, *Deutschland* was greeted with German–American bands playing the patriotic song, 'Die Wacht am Rhein' ('The Watch on the Rhine'). König and his crew were treated as heroes.

The British reaction to the voyage played right into the Germans' hand. Adm Patey deployed warships in patrols off the Chesapeake to catch *Deutschland* on her return voyage. American newspapers, skilfully orchestrated by Ambassador Bernstorff in Washington DC, condemned this latest British violation of US waters and the Freedom of the Seas.

On 2 August König sailed for Germany. *Deutschland* was carrying 348 tons of rubber (257 tons of which was carried externally), 341 tons of nickel and 93 tons of tin. The value of the rubber alone was more than $17,000,000 – several times more than the construction costs of *Deutschland* and her sister. On 24 August *Deutschland* arrived back in *Bremen* to a tumultuous welcome. *Deutschland* was to make one more successful voyage, this time to New London, Connecticut, but her sister-ship *Bremen* was sunk on her outward maiden voyage to Norfolk, Virginia.

The merchant U-boats had more than proved their viability. However, these operations came to an end on 1 February 1917 when the Germans launched their third and final campaign of unrestricted submarine warfare. *Deutschland* and her sister-ships were taken into the Navy and prepared for *Handelskrieg* on the high seas. Though they succeeded in their immediate aim of bringing vitally important raw materials back to Germany, the use of these large submarine was, in a way, short-sighted. A submarine can exert political influence out of all proportion to her size or value as a combatant – and *Deutschland* had shown just that. Her arrival in the United States had unleashed a good deal of pro-German feeling which was indicative that not all US public opinion favoured entry into the war on the side of the *Entente*. The reaction of the American press over the British deployment of patrols off the Chesapeake was another such indicator. A confrontation between *Deutschland* and a British warship in, or just outside, American waters would have had the most profound effect on American opinion. The German merchant U-boats were a potentially powerful force for at least keeping America out of the war.

The island of Malta provided British submarines with an opportunity for similar operations during the Second World War. Malta lay strategically half-way between the Italian mainland and the North African coast. It was therefore the ideal base, from which British cruisers,

A view of the submarine base at Malta in December 1941. Malta was the focus of a series of supply operations by British submarines which were known as the 'Magic Carpet' runs. (US Navy)

HMS Porpoise *participated in a number of 'Magic Carpet' operations to Malta. (Wright & Logan)*

destroyers, submarines and aircraft were to exact an increasingly heavy toll on Axis shipping. The 'siege of Malta' began the day after the Italians declared war on Britain and France on 10 June 1940 (the first air raid took place on 11 June) and would last until November 1942. The story of the various convoys which fought their way through to Malta against fierce opposition is one of the great episodes of the Second World War. However, the convoys were regarded as major fleet operations, requiring the despatch of units from the Home Fleet as well as Mediterranean-based forces. What was required was a discreet way of maintaining a regular supply to Malta – and this is where the submarines came in.

Without being distracted from their primary aim of sinking Axis shipping, British submarines participated in regular supply runs to Malta from June 1941 onwards. Every submarine sailing for Malta from Gibraltar or Alexandria would be expected to carry some items of stores. However, the main players in this game, which soon came to be known as the 'Magic Carpet', were the minelaying submarines *Rorqual* and *Cachalot*, the fleet submarine *Clyde* and the larger boats of the O, P and R classes. Strangely enough, the British had one submarine at their disposal whose potential for store-carrying they failed to appreciate. This was the monster 2,800-ton French *Surcouf* whose hangar would have made an cargo bay. But *Surcouf*'s services were never requested, perhaps owing to concerns about the reliability of her crew.

The Porpoise-class minelayers and *Clyde* proved particularly effective in these operations. These submarines had a large casing beneath which supplies could be stored and the minelayers in particular found that the

HMS Rorqual *in Grand Harbour, Malta, in February 1943. This photograph shows how the mine casing which ran the full length of the submarine on top of the pressure hull made these boats very suitable as store carriers. (US Navy)*

tunnel normally used for mine stowage proved an excellent cargo space. On one occasion *Rorqual* was able to carry 24 personnel, 147 bags of mail, 2 tons of medical stores, 62 tons of aviation spirit and 45 tons of kerosene. This cargo manifest indicates the main items which formed the bulk of submarine cargoes: aviation fuel; medical equipment and the all-important morale-booster for the garrison of the island: mail from the UK. Not mentioned in the official lists of cargo were the 'unofficial' items – cases of gin for the various wardrooms in Malta and on one occasion a consignment of records for Lord Gort, the island's Governor.

Cargo was stored inside the submarines in every conceivable space. *Clyde* actually had one main battery removed in order to create a 'hold' for store-carrying and which also provided an excellent venue for parties![3] In addition, containers were welded to her casing in which either submarine or aircraft torpedoes could be carried. Cargo could also be carried externally between the casing and the pressure hull or in the mine casing of the minelayers. This could cause problems, particularly when trimming the heavily laden boat. When on the surface the submarines had to maintain full buoyancy in order to protect the cargo from being damaged by the sea. In *Cachalot* wooden packing cases stowed in the casing absorbed so much water that the first lieutenant had to pump out 1,000 gallons of water from the internal tanks in order to compensate.

The carrying of petroleum products posed the greatest challenge. Aviation spirit or kerosene could either be carried in cans stowed beneath the casing (in July 1941 *Talisman* carried 5,500 gallons in this fashion) or 'in bulk' in the submarine's external fuel tanks. When cans of petrol were carried the submarine's diving depth was restricted 65ft, just below periscope depth, and then only for very short periods of time. Carrying fuel in the external tanks resulted in fumes venting in the normal way through the expansion system and posing a significant fire hazard. Smoking had to be forbidden on the bridge and the use of pyrotechnic recognition signals had to be prohibited.

Supply runs were made roughly once every twelve days and it is undeniable that without the 'Magic Carpet' Malta could not have remained in business as an effective base. Convoys could deliver a huge amount of bulk items such as flour and foodstuffs but submarines were able to fulfil the daily requirements of the ships, submarines and aircraft that made Malta such a thorn in the Axis side.

The main arena for cargo-carrying during the Second World War was between Europe and the Far East. Britain lost no time in implementing a blockade on Germany following the declaration of war in September 1939 and Germany was faced with almost the very same situation as twenty-five years before. However, there were a number of substantial differences between Germany's situation in 1914 and 1939. Japan was now her ally, with access to vast supplies of raw materials in Manchuquo, her Chinese client state. These resources would soon be extended by Japan's lightning war of conquest in South-East Asia in 1941–2. Russia was neutral, but following the Nazi–Soviet Pact of August 1939 was a benevolent neutral. Supplies from the Far East could and did reach Germany overland via the Trans-Siberian Railway.

This cosy arrangement was abruptly ended when Hitler invaded the USSR on 22 June 1941. The only route between Germany and the wealth of the Orient was now by sea. The Allied air and surface blockade made operations by surface blockade runners a hazardous affair (although a number did get through), so the submarine would be the preferred medium.

The Germans had first made use of submarines for transport duties during the Norwegian Campaign in 1940 after the horrendous toll on the merchant shipping exacted by British submarines and surface forces. *U29*, *U32*, *U43*, *U101*, *U122* and *UA* (the ex-Turkish *Batiray*) all took part in supply runs to Norway. A typical cargo, that carried by *U43*, consisted of 13 tons of infantry small arms ammunition and 15 tons of 2cm ammunition. The largest cargo was that taken by *UA*. It consisted of one complete 8.8cm gun, its mounting and accessories, 1,150 8.8cm shells, sixteen 2,500kg bombs, 165m^3 of aviation fuel, 5 tons of aircraft lubricating oil and 9 tons of engine lubricating oil for U-boats. For the rest of the war all German U-boats proceeding to Norway would carry stores. *U534*, a Type IXC/40 submarine, was sunk on 5 May 1940 while *en route* to Norway. When she was raised in 1993, she was found to be carrying tinned food, a significant quantity of engine parts and a large supply of condoms.

However, the impetus to build special transport U-boats came from a surprising direction. On 19 November 1942 Hitler announced his intention to plan an invasion of Iceland and the construction of an air base there. Transport submarines would be required to keep the base supplied

The Yeoman of Signals hoists Porpoise's *twin Jolly Rogers on her return to England in late 1942. The upper flag relates to her conventional submarine activities; the lower flag, on which the letters 'PCS' (Porpoise Carrier Service) can be seen relate to her 'Magic Carpet' runs to Malta. Four white bars can be seen on the flag. Each one represents a successful operation. (Author)*

with personnel, stores and equipment in the face of the British blockade. Whatever the Führer's concerns, however, it is clear that the *Seekriegsleitung* saw the role of transport submarines largely in terms of working between Europe and the Far East.

The first German design for a submarine transport was designated the Type XIX but was swiftly abandoned on account of the powerplant being untested. It was replaced by the Type XX, which used the successful powerplant fitted in the Type XIV *Milch Cow* tanker submarines. The hull form was quite boxy and followed the lines of the Type XB minelayers. It thus offered considerable opportunities for cargo stowage in containers stowed between the pressure hull and the casing. Estimates indicated that some 400 tons of rubber could be carried in this fashion. The keel could also be used for the stowage of tin in ingots, or molybdenum and tungsten in pewter or tin cases. Internal stowage allowed for approximately 220m^3 of cargo, items such as quinine, mica or oils. On 3 March 1943 thirty Type XX U-boats were ordered, fifteen from DE (Deutschewerke) in Hamburg Finkenwerder and fifteen from Bremer Vulkan. The first boat was to be complete by August 1944 and thereafter three boats per month were to be delivered. A condition of the contract, however, was that work on the Type XXs could not take the place of work on operational submarines.

The Type XXs would not be ready for over a year – and in the end wartime difficulties would mean that none of the class would ever be completed. But the Germans still needed a transport submarine. One solution lay in converting existing Type VII submarines for this purpose. Two schemes were put forward. The first involved turning the fore ends into a cargo compartment by the complete removal of the torpedo tubes and torpedo stowage racks. Such a conversion would enable 105 tons of cargo to be carried: 25 tons internally, 35 tons externally between casing and pressure hull and 45 tons in the keel. A second option envisaged the same modifications as above but also enlarging the saddle tanks. This would give a carrying capacity of 250 tons. While both conversions were practical it was recognised that the withdrawal of existing *Fronteboote* for conversion to transport duties was a measure of last resort.

Another alternative was to request the transfer or loan of submarines from the Italians, who were already making extensive use of their submarine fleet for transport duties to and from North Africa. The Italians were facing the same economic position as the Germans and had already ordered twelve submarines of the Romolo class (with a displacement of 2,155/2,560 tons and a cargo-carrying capacity of 600 tons) specifically for transport duties to and from South-East Asia.

Dönitz first asked the *Supamarina* whether the ten Italian submarines then based at Bordeaux could be converted for cargo-carrying. The boats were large and had not proved themselves suitable for convoy battles in the North Atlantic. The idea had some merits: only six weeks would be required for the conversion. However, the cargo-carrying capacity of all ten boats would be a mere 1,500 tons and the submarines lacked the endurance to make the passage to the Far East unaided. Accordingly, a tanker would have to be stationed somewhere in the remote South Atlantic for both the outward and return passage. On 18 March the

Italians consented, with the proviso that the most modern of their submarines at Bordeaux, *Cagni*, be exempted. The price of the transfer was to be the release of ten Type VIIC boats to the *Regia Marina*. In this example of Axis relations the Italians definitely got the better bargain! The conversion of the Italian boats was carried out under German supervision with the result that by May 1943 the first three boats were ready to leave Bordeaux. They were *Tazzoli*, *Giuliani* and *Cappellini*. To disguise the purpose of the operation the boats were given the cover name *Aquila* and the first three boats became *Aquila I, II* and *III* respectively. In June 1943 *Barbarigo* (*Aquila V*) and *Torelli* (*Aquila VI*) were despatched. The German tanker *Burgenland* was despatched to the South Atlantic to refuel the boats but of the five only *Giuliani*, *Cappellini* and *Torelli* arrived. A sixth submarine, *Bagnoli*, was despatched after the Italian armistice, having been renamed *UIT22*, but was sunk off the Cape of Good Hope.

While the conversions were going ahead, Dönitz attempted to lay his hands on the dedicated Italian Romolo-class transport submarines. Adm Riccardi, Chief of *Supamarina*, refused even though, since the Axis surrender in North Africa, he had no further use for these submarines. Only two of the class, *Remo* and *Romolo*, were ultimately completed. *Remo* was torpedoed by HMS *United* on 15 July 1943, and *Romolo* was sunk in an air attack off Augusta on 18 July 1943. The remainder were seized by the Germans in September 1943 while under construction. All were either bombed while fitting out or broken up on the slip after the war.

Having made the long journey to Singapore, the three surviving Italian boats were almost instantly impounded by the Japanese on 18 July 1943 following the fall of Mussolini. It took some labyrinthine negotiations on the part of the Germans to have the boats released to them and it was not until the beginning of 1944 that *Cappellini* (now known as *UIT24*) was ready to sail for Europe with a cargo of rubber, quinine and precious ores. However, *Cappellini* failed to rendezvous with the tanker *Brake* so her commanding officer broke off the voyage and returned to Penang. *UIT23* (ex-*Giuliani*) was no more fortunate, being torpedoed and sunk by HMS *Tally-Ho!* in the Malacca Straits on 15 February 1944. The *Seekriegsleitung* then abandoned the idea of using the Italian boats as cargo transports. The two survivors, *Cappellini* and *Torelli*, were eventually employed as short-sea transports between the German bases in Malaysia and Borneo, and the Japanese mainland.

The last straw, the Germans now began to consider converting the operational U-boats serving in the Far East to carry cargo on their return voyage to Germany. In June 1943 nine Type IXC and two Type IXD boats had been despatched to the Indian Ocean to operate from the port of Penang. It was estimated that a Type IXC could carry 145 tons of cargo: 115 tons of tin (of which 100 would be stowed in the keel), 10 tens of molybdenum, 9 tons of tungsten, 10 tons of rubber, 0.5 tons of quinine, and 0.2 tons of opium. The larger Type IXD could carry 120 tons of tin, 15 tons of molybdenum, 80 tons of rubber, 1 ton of quinine and 0.2 tons of opium – a larger total of 215 tons. In both the Type IXC and IXD conversions the bulk of the tin would be carried in the keel, and the rubber would be stowed between the casing and the pressure hull. Other

precious metals were soldered into tin containers and stowed in the bilges or the torpedo tubes or the fore ends.

The first U-boat to return was *U178* on 24 May 1944 followed by *U188* in June 1944. These two boats were followed by *U843* and *U861* which reached Norway in April 1945 carrying 100 tons of zinc apiece (*U843* was sunk on 8 April 1945 while proceeding to Germany). *U510* was at sea when the German surrender took place and she put into St Nazaire in France. Lastly, *U532* surfaced off the west coast of Ireland to receive the German surrender signal, and was subsequently escorted into Liverpool where some of her cargo of 100 tons of tin, 60 tons of rubber, 8 tons of wolfram and 5 tons of molybdenum were laid out on the quayside for the curious to inspect.

The German attempt to bring raw materials from the Far East was a failure. Only the economic stranglehold on Germany established by the Allies caused so much effort and expenditure to be made on a project for which the submarines involved were so ill-suited. They could never carry sufficient raw materials to keep Germany's war economy fully supplied. As one example, the Armaments Ministry informed the *Seekriegsleitung* that

1,000 tons *per month* of cargo from the Far East was required to meet production requirements.

Perhaps the strangest supply voyage of all concerns *U234*, a Type XB minelayer. On the other side of the world the Japanese had an equal, but less pressing, interest in maintaining a link with Europe. Japan's greater concern was with the acquisition of technical equipment from Germany, rather than of raw materials. *I.30* (Cdr Sinobu Endo IJN) was the first submarine to make the voyage to Lorient in 1942. After loading with stores she returned to Japan and arrived safely in Singapore in October 1942. Disastrously, she was mined on 13 October when leaving Singapore for Japan. She sank with the loss of 87 of her crew of 100 and all the German equipment.

The next boat to sail, *I.8* (Cdr Tetsunoke Ariizumi IJN), left Penang on 6 July 1943. Off the Azores she rendezvoused with a German U-boat and was fitted with a radar-warning receiver to detect Allied search radars. On 5 September *I.8* made a triumphant entry into Brest after a 61-day voyage. After storing with supplies, *I.8* began the return voyage and was back in Japanese waters in December 1943. She was the only Japanese submarine to complete the 30,000-mile round trip between Japan and France. The third submarine suffered the fate of *I.30*. After making the successful round trip *I.29* was torpedoed south of Formosa by USS *Sawfish* on 26 July 1944. The fourth, *I.34*, got no further than Penang on her outward trip before being sunk on 13 November 1944 by the British submarine *Taurus*. The last Japanese submarine to attempt the voyage, *I.52*, was returning from Europe when she was attacked and sunk by aircraft from USS *Bogue* on 24 June 1944.

A more pressing requirement for the Japanese was the need to supply their far-flung ocean possessions in the face of an American offensive that gathered in speed and ferocity month by month. In November 1942 all available I-class submarines were withdrawn from operations and concentrated at Rabual on the island of New Britain. From there they would transport supplies to the Japanese garrisons in the Solomon Islands and especially Guadalcanal. The aircraft-carrying submarines were found to be the most useful since over 50 tons of food and ammunition could be stowed in the hangar – a two-day supply for the Guadalcanal garrison.

These *Mogura* (Mule) operations were deeply distasteful to Japanese submariners. They ill-accorded with the noble concept of Bushido. The first operation was by *I-176* (Lt Cdr Yahachi Tanabe IJN) and by February 1943 some 1,500 tons of supplies had been taken to Guadalcanal in twenty-eight supply runs. As the American offensive gained ground these operations were extended to New Guinea and the Aleutians. Up to September 1943 over 3,500 troops were ferried into New Guinea by submarine. During the remainder of the Pacific War, Japanese submarines carried out supply runs to all parts of the island empire. Submarines employed in these operations were modified for their new role by having deck-guns removed and as much internal space as possible converted for cargo stowage.

Provisions for the ship's company were cut to the bare minimum in order to increase cargo stowage. Buoyant rubber sacks of rice and oil drums could be carried between the casing and the pressure hull and either floated

Opposite: *Japanese I-class submarines were used to transport technical equipment from Germany to Japan and later to supply the beleaguered Japanese forces in the Pacific in the face of American gains. These I-class boats are anchored in Tokyo Bay after the surrender in 1945. (US Navy)*

ashore or transferred to a small boat. The Japanese were nothing if not ingenious and even experimented with firing buoyant sacks of rice from a torpedo tube! Other methods included the development of the *Unpoto* powered lighter which could carry artillery ashore and the *Unkato* dumb stores container, with a capacity of 20 tons, which was towed behind the submarine – with difficulty as it was found out. The *Unkato* could either be released to drift ashore or was towed ashore by small boats.

A secondary priority was evacuation. Contrary to popular belief, Japanese forces did not always make last ditch stands in the face of the Americans. The Japanese were keen to evacuate key personnel from islands facing invasion, especially technicians – of which Japan had a great shortage – and staff officers, and even valuable items of equipment. The most important 'evacuees' were aircrew. Any pilots shot down over Japanese-held territory would be guaranteed a place on the next homeward-bound submarine. The largest single evacuation was from the island of Kiska in the Aleutians when a force of fifteen Japanese submarines removed the garrison of 820 men.

Supply operations using submarines were the subject of fierce argument within the Japanese High Command where Army–Navy relations were never better than mildly poisonous. To put it in a nutshell, the Navy objected violently to being asked to run a cargo service for the Army. Arguably, the Navy had a point: this was not what the boats had been built for and these operations sapped the operational readiness of the crews. On the other hand, the Army contended that the Japanese submarine fleet's contribution to the war effort was hardly significant and if the submarines were not sinking ships they may as well be carrying rice. The Army also had a point – and their arguments were sharpened by the reality of combat: Japanese soldiers were fighting and dying, hampered by a critical lack of supplies.

The Navy eventually had to give in to Army demands and the result was the design and construction of specialised transport submarines. Thirteen submarines of the D1/D2 type and twelve submarines of the SS/HA-101 series were built. The D1/D2 type were the only Japanese submarines not fitted with torpedo tubes. The D1 design of 1440/1779 tons provided for the carrying of 22 tons of cargo internally, plus 110 troops. The cargo compartments were forward, where the fore-ends would otherwise be, and aft of the control-room. The after compartment had an endless conveyor belt sloping upward to the cargo hatch. Externally another 60 tons of cargo could be carried on the casing aft of the conning tower, along with two 42ft landing craft. Of the twelve D1s only *I-363*, *I-366*, *I-367* and *I-369* survived the war. The one D2 boat, *I-373*, which was a slightly larger design, was sunk in April 1945. Interestingly enough, both types were fitted with a schnorkel – proof that the Japanese respected the efficiency of American surface and air-to-surface radar. The heavy losses in the D1/D2 type resulted in the construction of the smaller SS/HA-101 type. These boats were of 370/420 tons displacement and could carry only 60 tons of cargo. All ten boats that were completed survived the war.

The Japanese Navy's unenthusiastic reaction to supply operations caused the Army to design and develop their own submarines. The Navy were quite keen on the idea since it would free their submarines from this tedious task and even went so far as to offer technical advice on the Army boats' construction. The offer was declined. Twelve Army submarines of the YU-1 class and fourteen boats of the YU-1001 class were built: both types were capable of carrying 40 tons of cargo. A third Army class, the YU-2001, was projected but never completed. The designs show a strong German influence in the shape of the conning tower, possibly indicating that the Army had German help with the design. This must be the only occasion where an Army has built its own submarines out of frustration with the attitude to its requirements shown by a sister service.

Cargo-carrying as a role for submarines featured in the US Navy's postwar plans. Reference has already been made to the SSP submarine transport conversion in the chapter on special operations. Following the two SSP conversions, USS *Barbero* (*SS317*) was converted as a cargo-carrier. Cargo compartments with large deck hatches were installed in the former after torpedo-room, forward engine-room and part of the after battery compartment. The enlisted men's accommodation was moved to the forward battery compartment while the wardroom was moved aft. Two of the normal fuel oil tanks were modified to carry gasoline, and new motors were fitted together with a schnorkel. However, after operational trials the Navy decided to take the concept of a submarine freighter no further and *Barbero* joined the Regulus programme.

Undoubtedly the most hazardous point for a submarine engaged in cargo-carrying or blockade-breaking comes as she nears her destination and must surface. The long passage may mean that the boat's position does not accord with the navigator's calculation, and the submarine must also face the problem of establishing her identity with friendly forces who may be inclined to shoot first. Furthermore, surfacing exposes the submarine to all the risks of detection and attack by enemy forces. This is the fate that befell the Argentine submarine *Santa Fe* in May 1982. *Santa Fe* had sailed from Mar del Plata in Argentina bound for South Georgia with stores and supplies. On approaching the island she surfaced and transmitted an arrival signal which was instantly picked up by the British. The rest of the story is well known. *Santa Fe* was repeatedly attacked by Royal Navy helicopters with missiles and machine-guns and limped into Grytviken badly damaged and on fire.

Cargo-carrying submarines like *Deutschland* were initially successful because of the lack of opposition deployed against them but operations in the Second World War showed the inherent dangers and disadvantages of using submarines as transports. Nevertheless, military requirements have usually always given way before political or economic necessity and there is no reason why this should not be so in the future.

NOTES

1 Britain sought to classify as many goods and materials as possible as either Absolute or Conditional contraband, which meant that they could legally be seized. See Kemp, P., *Convoy Protection: The Defence of Seaborne Trade* (London, Arms & Armour Press, 1992).
2 PRO ADM137/2918. C-in-C North America and West Indies to Admiralty, 26 June 1916.
3 Cdr John Bull to author, 12 February 1988.

KEEPING AN EYE
ON THE ENEMY

Surveillance is a task tailor-made for the submarine, particularly the nuclear submarine, which is freed from the constraints of having to surface to charge the battery. In wartime, submarines provide an ideal means of keeping an eye on the movements of enemy units. During the Second World War the operations of British submarines in home waters (from March 1942 until November 1944) were almost exclusively dominated by the need to throw a screen around the various Norwegian fjords to intercept the German battleship *Tirpitz*. The *Tirpitz* example shows some of the difficulties associated with using a submarine for surveillance in wartime, namely that the submarine might find it difficult to report a sighting for fear of detection or direction finding, and that the submarine lacks the speed to exploit the contact.

On 5 July 1942 the German battleship *Tirpitz* sailed from Altafjord to attack the ships of convoy PQ.17. HMS *Unshaken* (Lt C.E. Oxborrow RN) made contact at 1922/5 in position 71°40'N 28°19'E. She first sighted smoke and increased speed to close the range but was put down by an aircraft. By 20.13 hrs the unmistakable superstructure of *Tirpitz* was in view with *Hipper* astern of her. *Tirpitz* was steering a course of 080°, bearing 200°. Oxborrow would only get a chance if *Tirpitz* altered course to the north. By 20.20 hrs the range had opened and Oxborrow reluctantly broke off his attack. His next duty was to report his sighting. He surfaced at 21.18 hrs but was immediately put down by an aircraft. Finally, at 21.57 hrs he was able to surface and transmit unmolested.

If not able to attack themselves, submarines could shadow an opponent, reporting the speed, course and composition of the enemy force so that other units can either clear the area or reinforcements be summoned. At 20.10 hrs on 6 April 1945 USS *Threadfin* sighted the Japanese battleship *Yamato*, escorted by the cruiser *Yahagi* and eight destroyers, emerging from the Bungo Strait. The range was too great for an attack but *Threadfin* maintained pursuit until 23.45 hrs by which time she had successfully established the Japanese force's speed, course and composition. Her reports allowed the Americans to make the appropriate dispositions with the result that *Yamato*, *Yahagi* and four destroyers were sunk by carrier-borne aircraft the next day.

In peacetime the use of submarines for surveillance is another matter. The passage of warships through another nation's territorial waters is governed by a series of international conventions. Permission should be

HMS Odin *was one of the large submarines of the O, P and R classes designed and built for service in the Far East. Submarines of this type carried out covert intelligence-gathering operations in Japanese waters between 1939 and 1941. (Author)*

requested, unless the vessel concerned is driven into territorial waters by bad weather or distress, and once inside territorial waters warships should not make use of their radio equipment. However, a submarine can enter another nation's territorial waters with impunity – the only rule in force being the unwritten Eleventh Commandment 'Thou shalt not get caught'. The first naval power to make deliberate use of the submarine for covert surveillance purposes was Britain between the wars in the Far East.

In 1922 the British government had made a dubious bargain, sacrificing an alliance with the Japanese government that protected British interests in the Far East in favour of engaging in an arms-control system, in an attempt to limit naval armaments. Britain's relationship with Japan, on which the security of Britain's eastern empire largely depended, deteriorated rapidly. There was no possibility of Britain maintaining a separate battle fleet in eastern waters so the submarines of the 4th Flotilla became the main striking force of the Royal Navy in the Far East. Working from supposedly impregnable bases at Hong Kong and Singapore, British submarines could range into Japanese waters to blockade Japanese bases and attack their powerful battle fleet. Part of the preparations for any war involves gathering as much intelligence as possible about the enemy, yet in this case Japanese secrecy about their navy represented nothing less than an information black-out. Accordingly, British submarines were despatched in search of definite information.

Low-level or limited conflicts also provide submarines with unrivalled opportunities for surveillance or intelligence-gathering without upping the political or military temperature. The 1982 Anglo-Argentine conflict over the Falkland Islands was never formally categorised as a war and in the early stages there was a chance that diplomatic negotiation might bring about a solution. Nevertheless, the British SSNs, which had been ordered to the South Atlantic ahead of the Argentine invasion, were able to signal much useful intelligence to London. On 15 and 17 April HMS *Spartan* reported that the Argentinians were laying mines off Port Stanley. The nuclear submarine, using her ability to spend long periods submerged, was able to report on enemy dispositions while remaining unobserved and avoiding confrontations that might escalate the crisis.

It was the period of super-power rivalry after the Second World War that really allowed the submarine to explore its full potential as an intelligence-gatherer. The often tense relations between America and the USSR made it all the more imperative for each side to gather as much information as possible about the other side's equipment and performance. At the same time, the electronic revolution which was transforming naval warfare was increasing the amount of intelligence available for the taking. The periscope observations which had been the lot of British submariners in the Sea of Japan before 1941 could now be supplemented by the interception and recording of enemy radio traffic and radar signatures and the acoustic recording of enemy ships and submarines. Lastly there was the supreme prize of detecting and tracking an SSBN on the way to or returning from patrol.

Details of such operations are notoriously hard to come by – no government is prepared to admit to a flagrant breach of international law by admitting that its submarines have been conducting covert operations within another country's territorial waters. However, it is safe to assume that Britain, Russia and France have sent their SSNs to operate in Russian waters and vice versa. Details about operations by western boats are easier to obtain – one of the aspects of living in a democracy is that governments are less able to control the activities of either over-inquisitive journalists or garrulous submariners. Britain's first covert patrol in Russian waters in 1953, by the T-class submarine HMS *Totem* (Cdr John Coote RN), was described by her CO as 'batting against really serious opposition'.[1] The patrol was subsequently nearly compromised by a junior rating who was so thrilled by the experience that he gave a full account of *Totem*'s exploits to the other occupants of the railway carriage in which he was going on leave! The rating was dealt with by the authorities – presumably the Russian authorities did not have such a problem.

In May 1975 the *New York Times* revealed details of the US Navy's 'Holystone' intelligence-gathering programme whereby US submarines had made frequent patrols inside Russian waters since the early 1960s including patrols well within the three-mile limit. Representatives of the newspaper were shown a series of photographs of the hull of a Russian Echo II-class SSGN reportedly taken inside Vladivostock harbour. These operations were not without risk. In December 1967 the USS *Ronquil* (SS396) reportedly narrowly avoided capture by Russian naval forces while engaged in a Holystone intelligence-gathering mission. The submarine caught fire near the Russian coast while dived and was surrounded by Russian destroyers which attempted to force her to surface. The *Ronquil* eluded the Russian ships and escaped to safety. On another occasion an unknown American SSK was forced to surface by Russian forces and transmitted a request for assistance *en clair*.

In December 1969 reports emerged that the USS *Gato* (SSN615) had collided with a Russian submarine on the night of 14/15 November, at the entrance of the White Sea in the Barents Sea. A crew member is quoted as saying that the *Gato* was struck in the heavy plating that serves as a protective shield around the nuclear reactor, but the ship sustained no serious damage. However, the ship's weapons officer immediately ran down two decks and prepared for orders to fire a nuclear-armed SUBROC anti-submarine missile and three nuclear-armed torpedoes. The accident reportedly occurred during a Holystone operation. According to former *Gato* crew members their commanding officer was ordered to prepare false reports showing the submarine had suffered a breakdown and halted its patrol two days prior to the collision. The *Gato*'s commanding officer refused to comment when he was contacted due to security reasons.

In May 1974 the USS *Pintado* (SSN672) reportedly collided almost head-on with a Russian Yankee-class SSBN while cruising 200ft deep in the approaches to the Petropavlovsk naval base on the Kamchatka Peninsula. The Russian submarine surfaced immediately, but the extent of the damage was not known. The *Pintado* departed from the area at top underwater speed and proceeded to Guam where she entered dry-dock for repairs lasting seven weeks. The collision smashed *Pintado*'s active sonar, a starboard side torpedo hatch was jammed shut and the starboard diving planes received moderate damage.[2]

The most perplexing incident involving an American submarine concerned the USS *Baton Rouge* (SSN689). On 11 February 1992 the

Damage to a Russian Delta-class SSBN following a collision with an American SSN. (Author)

HMS Sceptre *ploughing through the waves in the Firth of Clyde, January 1980.*

Baton Rouge was on patrol off the Kola Inlet, where is located the main Russian submarine base at Severomorsk. She is reported to have encountered, manoeuvred around and then collided with the Russian *Barracuda*, a Sierra-class SSN. Details of the damage suffered by both boats were not made public but the incident led to some heated diplomatic exchanges. Some indication of the damage suffered by *Baton Rouge* can be gauged by the fact that she was subsequently paid off. Apparently the cost of repairs was too great. She had been in service for only two years and remains the only SSN688 class boat to have been scrapped.

British submarines have not been immune from such incidents. In 1981 HMS *Sceptre* collided with a Soviet nuclear submarine and in January 1987 it was reported that HMS *Splendid* had lost her towed array in a collision with a Soviet Typhoon-class PLARB on Christmas Eve 1986. It was unclear whether the collision was solely the result of manoeuvring in close quarters or whether the Russians were deliberately trying to obtain this sensitive technology.

Russian submarines perform the same operations, though details of their incursions into British or American waters have not been revealed. It is safe to assume that some Russian submariners are as familiar with the waters around the Clyde, the Isle de Longue off Brest (the base for French

SSBNs) and King's Bay, Georgia as they are with those of the Kola Inlet. The Swedish government, however, has made no secret of their suspicions about Russian submarine movements in their waters and have adopted a tough response, including the dropping of live ordnance on suspected intruders.

In October 1982 it fell to the Russians to suffer the most embarrassing case of being caught *in flagrante*, when a Whiskey-class SSK ran aground 10 miles from the Swedish naval base of Karlskrona, 300 miles south of Stockholm, well within Swedish territorial waters. The Swedish government alleged that the submarine was engaged in illegal reconnaissance and that there was good reason to believe that the vessel was carrying nuclear weapons. The Swedes demanded both an apology and an explanation. When the submarine captain was questioned he contended that bad weather and a faulty compass led to his straying into Swedish waters. The Swedes dismissed this claim with the assertion that his navigation must have been excellent to have made such a deep incursion into Swedish waters and coincidentally so near to a major Swedish naval base. The embarrassment of the Russian government was acute and the Swedes had no intention of letting them off lightly. No wonder the submarine's CO was seen making throat-cutting gestures – a reference to his own likely fate – to Swedish journalists circling the boat.

On 29 October 1982 a Russian tug sent to retrieve the submarine was turned back by Swedish warships. At the same time another unidentified submarine, presumably Russian, was spotted within Swedish waters and was pursued by Swedish warships and helicopters until she returned to international waters. Eventually the Swedes refloated the submarine themselves: the approach of bad weather meant that the submarine would probably sustain further damage by being worked against the rocks.

The Swedes extracted the maximum diplomatic advantage from the situation. Swedish naval officers and environmental groups circled the submarine with Geiger counters to take radiation readings which provided sufficient evidence for the Swedes to announce on 5 November that the submarine had nuclear weapons aboard. Foreign Minister Ullsten said that, 'it must be very embarrassing' to have this information released when the Russians 'have created the impression that they are more in favour than the United States' of arms control. On 6 November the submarine was finally returned to the Russians. It was later reported that the Russian government had agreed to pay the Swedish government $212,000 for costs arising from the incident.

Shadowing the opposition's fleet during ocean manoeuvres is another task submarines are required to perform in peacetime. The amount of intelligence that can be gleaned in this fashion is considerable, but it exposes the submarine to significant risks and has serious political implications. A submarine can choose to remain undetected while performing this task, but there are occasions when revealing her presence can pay dividends. If sonar conditions are poor and have resulted in contact with the surface forces being broken, the submarine may choose to use her radar to relocate them. The response of, say, a NATO task force on detecting the distinctive transmission of the Snoop Tray radar used by Russian Victor III SSNs (the type most commonly employed on surveillance operations) could be instructive. Similarly should a British Trafalgar-class SSN proceed through a Russian submarine vs. submarine exercise area with her active sonar booming away, the resulting reaction on the part of the Russian submarines would be well worth recording.

However, there are serious risks in these operations. Intelligence-gathering, either by visual or electronic means, requires that the submarine remains fairly close to the 'target'. There is therefore the risk of collision when manoeuvring close to a surface ship or a task group (or for that matter a submerged submarine). In 1984 the American aircraft carrier *Kitty Hawk* (CV-63) was participating in exercise 'Team Spirit 1984' – joint USN–South Korean manoeuvres. While proceeding towards the Yellow Sea through the southern Sea of Japan, *Kitty Hawk* collided with a surfacing Russian Victor I SSN. *Kitty Hawk* sustained a minor hole below the waterline in an aircraft fuel tank on the starboard side but was able to proceed with the exercise. The Russian vessel was observed lying dead in the water for a while with a large gash across the after casing. Polite offers of assistance from *Kitty Hawk* and other US ships in company were declined and the submarine continued to lie dead in the water until the Kara-class cruiser *Petropavlovsk* appeared, followed by a tug which towed the submarine to Vladivostock naval base. The Victor had been following the US ships for some days despite sustained efforts by the Americans to break contact. It appeared that the submarine was coming to periscope depth for a visual sight of the US ships when she collided with the carrier.

Other problems associated with shadowing an exercise are that the participating ships might mistake the shadowing submarine for one of their own and thus subject her to a good deal more attention than might otherwise be the case. This situation can yield a cornucopia of intelligence – but at some risk. Even worse is for a shadowing submarine to stray accidentally into a live firing area where 'hot' ASW ordnance is being tested. In times of international tension, the presence of a potentially hostile submarine in an exercise area can be a destabilising factor with the actions of both sides easily open to misinterpretation.

Another area of surveillance in which submarines have been involved is that of radar pickets, to provide advance warning of air or missile attack to either a surface task force or the national government. The Americans were the first to develop this role for submarines and their decision was heavily influenced by their experience of Japanese *kamikaze* air attacks, particularly in the spring and summer of 1945 off Okinawa. Converted destroyers had suffered heavily in this role and it was considered that a submarine picket offered a greater chance of both providing advance warning of an attack and being able to submerge and evade detection. The first two such conversions *Requin* (SS481) and *Spinax* (SS489) were put into service in 1946. Neither boat was formally classified as an SSR. Speed was of the essence during the conversion as the US planners were still contemplating the invasion of the Japanese home islands and so the radar equipment fitted was a mismatch of sets currently in service, and installed in a haphazard fashion through the boat. Naturally, so many problems developed with these boats that the USN instituted the 'Migraine' programme under which three 'purpose-built' radar picket submarines were produced.

The Migraine I conversions involved the submarines *Tigrone* (SS419) and *Burrfish* (SS312). The after torpedo-room was stripped of the two torpedo tubes and converted to an accommodation space while the former crew's mess became the aircraft control centre. Two of the forward torpedo tubes were also removed in order to provide more berthing space. The radar equipment fitted consisted of SS, SV-1, SV-2 and AN/BPS-2 radars together with a YE-2 homing beacon. In the two Migraine II conversions (*Spinax* and *Requin*) the air control centre was placed in the after torpedo-room from which the torpedo tubes had been removed. However, in the Migraine II conversions, the SV-2 height-finding radar was placed on a low mount on the after casing instead of on a pedestal mount as in the Migraine I boats. Otherwise the radar equipment was the same except for the substitution of SR-2 for the AN/BPS-2 set and a YE-3 homing beacon.

Both Migraine I and II suffered from a common problem – overcrowding, both of personnel and equipment. The crew of a Migraine conversion SSR was around 100 officers and men – a 25 per cent increase on a fleet boat's wartime complement. When combined with the installation of so much electronic equipment, it made for uncomfortable living. Accordingly in the Migraine III programme six thin-hull Gato-class fleet boats, *Pompon* (SS267), *Rasher* (SS269), *Raton* (SS270), *Ray* (SS271), *Redfin* (SS272)and *Rock* (SS274) had a 24ft section inserted into their hulls between the forward battery compartment and the control-room, in order to accommodate the air control centre and electronic equipment. Even so, the after torpedo-room was still converted to an accommodation space. Dedicated radar sets, the AN/BPS-2, AN/BPS-3 and AN/BPS-4, which permitted the accurate height-finding and tracking of aircraft, replaced the improvised models used in earlier conversions. To aid streamlining, the aerials were now carried on a streamlined sail instead of being mounted awkwardly around the conning tower. A lot of effort had gone into the radar picket conversions but it all came to naught when the US Navy announced in 1959 that it was abandoning the concept of radar pickets.

The Russians undertook similar radar picket conversions when a number of Whiskey-class submarines were fitted to carry radar at the after end of the conning tower. The conversions were known as 'Whiskey Canvas Bag' after the crude canvas structure which covered the radar's aerial from prying eyes. The Russian boats may also have had a subsidiary role of providing mid-course guidance for missiles fired from a ship, submarine or aircraft.

The post-Cold War international situation is deeply unsettling. In the place of super-power rivalry, there are a number of low-level regional conflicts which, in one way or another, involve American, European or Russian interests. At the same time the submarine-owning base is widening – the only weapons export that has actually increased over the last five years is the submarine[3] – as many countries achieve regional power status. There are now over 400 submarines in service throughout the world operated by a total of 43 nations. Some of these nations (for example, Iran and North Korea) present significant security concerns over their intentions, links with terrorist organisations and nuclear capability.

Surveillance on the submarines of these countries will have to be conducted on a much wider basis to build up data on their acoustic noise signatures. In 1982 the Royal Navy was critically handicapped in the South Atlantic by its lack of information on the acoustic signatures of the two Argentine Type 209s. The possibility that the humble torpedo might well become the carrier for a nuclear weapon means that surveillance of the growing Third World submarine fleet is likely to remain a continuing priority for some time.

NOTES

1 Capt John Coote to author, 17 May 1989.
2 Details of *Pintado*'s exploits and those of other US and British submarines are contained in a Greenpeace document on submarine operations and accidents available on the Internet.
3 Fitzgerald, Vice Adm, J.R., 'About Anti-Submarine Warfare', *The Submarine Review*, April 1997, p. 7.

MIDGET SUBMARINES

Theirs was a courage which is not the property, or the tradition or the heritage of any one nation. It is the courage shared by the brave men of our own countries as well as of the enemy and, however horrible war and its results may be, it is courage which is recognised and universally admired.
Rear Adm Stuart Muirhead-Gould on the Japanese midget submarine crews killed in the attack on Sydney harbour.

The midget submarine is one of the most potent weapons of war developed in the twentieth century. Yet it is an extremely old form of naval warfare with the first, although unsuccessful, attack being launched in 1776. In the early days of submarine warfare in this century, all submarines were 'midget' craft. But as the submarine grew, there arose a requirement for small craft which could penetrate a defended harbour and attack shipping. This single requirement has since been expanded to include a host of other roles. Experience with the midget submarine showed that these small craft could accomplish operations of considerable strategic importance with effects out of all proportion to the small size of these craft.

It was the stalemate in the Adriatic during the First World War which gave the stimulus for the development of these weapons. Neither the Italian nor the Austrian fleet would venture out from behind their protective nets and minefields, so the Italians sought ways of attacking the Austrian ships in their bases. The Grillo 'climbing tank' was one such (unsuccessful) weapon, but the *Mignatta* achieved its aim, and pointed a finger to the future, by sinking the Austrian battleship *Viribus Unitis* on 1 November 1918.

During the Second World War all the major belligerent navies, with the exception of the United States, France and the USSR, employed midget submarines or specially trained assault frogmen. The absence of France from this field is easily explained – her capitulation in June 1940 effectively removed the French from the war. The United States possessed conventional forces in abundance and thus did not need to resort to this form of warfare. However, the absence of the Soviet Union from this area of operations is puzzling given the pioneering work done by Russian engineers in submarine development. But the highly individualistic nature of midget submarine operations is not one that sat easily alongside the centralised Soviet command structure.

An Italian B-class submarine fitted with tracks to enable it to climb over net obstacles, Venice, 1917. (Author)

Three kinds of midget submarine made their appearance during the Second World War: human torpedoes (the Italian *Maiale* and British Chariot); small submersibles (the German *Neger* and associated craft and the Japanese *Kaiten*) and true midget submarines (the Japanese *Ko-Hyoteki*, the Italian CA/CB types, the British X-Craft and the German *Seehund*). These craft can also be further divided into the practical and therefore successful (British X-Craft, Italian *Maiale*); those that were enthusiastically designed but impractical (British Chariot, Japanese *Ko-Hyoteki*, German *Biber*) and the suicidal either by accident or design (British Welman, German *Neger*, Japanese *Kaiten* and its various derivatives).

It was the Italians who led the way. The *Regia Marina* was the only major navy to possess a unit, the Decima Mas, dedicated to special operations. The *Mignatta* had evolved into the *Maiale*, a two-man human torpedo, which was used to such deadly effect at Alexandria and Gibraltar. The activities of the Italian CA/CB midget submarines are less well known but worthy of attention, particularly the plan to attack shipping in New York – an operation which would have had the most serious effects in America but which was cancelled on the Italian armistice.

Japan was another of the early pioneers in this field. Before the war the Japanese developed the excellent two-man *Ko-Hyoteki*, an extremely advanced midget submarine. Japanese war plans concentrated on the great battleship engagements between the American and Japanese fleets which would decide the course of the war. To whittle down the American superiority in capital ships, *Ko-Hyoteki* were to launch mass torpedo

An Italian CB-type midget submarine at Costanza on the Black Sea in the winter of 1941/2. The CBs were useful little craft armed with two torpedoes and proved successful in the Black Sea. However, they were not employed to their potential in the Mediterranean. (Dott Achille Rastelli)

The 'well' made in the forward casing of the Italian submarine Leonardo da Vinci *in which she would carry the midget submarine CA.2 for an attack on New York. The attack never took place on account of the Italian armistice and remains one of the great 'what ifs' of the Second World War. (Dott Achille Rastelli)*

The graving dock at Kure in Japan in 1945 packed with Koryu *submersible craft. The Japanese hoped that sheer weight of numbers of these craft would defeat the American invasion forces.*

attacks. It was an ingenious idea and may well have worked. However, the Japanese decision to destroy the US fleet by a carrier strike removed their *raison d'être*. Instead they were employed in harbour penetration – a task for which they were not suited and at which they were less than successful. As the tide of the war turned against them, the Japanese resorted to suicide weapons such as the *Kaiten* and *Kairyu*. These were intended to overwhelm the Americans by sheer weight of numbers, but once deployed these craft proved no match for the range of anti-submarine measures employed by the Americans.

It was the Italian activities in the Mediterranean which spurred the British into the field. Britain had traditionally made no attempt to develop this sort of weapon – since the Royal Navy enjoyed a pre-eminent position in the world there was no need. It was only the need to attack the German battleship *Tirpitz* which pushed a reluctant Admiralty in this direction. The British initially copied the Italian two-man human torpedo and produced the Chariot, but these craft proved unsuccessful and never justified the time and resources devoted to them. Far more successful was the X-Craft, a potent four-man midget submarine which could be put to a variety of uses and was a most potent weapon of war. At the other end of the scale was the Welman, a useless craft whose design shows the effects of allowing enthusiasm to triumph over practicalities. British midgets saw

A British two-man Chariot human torpedo. This weapon was almost a direct copy of the immensely successful Italian Maiale. *(Author)*

action in all three theatres of war and their most significant success was the crippling of the German battleship *Tirpitz* in September 1943.

The Germans were the last into this field. While the U-boats were scoring significant successes in the Atlantic, the *Kriegsmarine* showed no interest in these craft. It was only when the Germans were faced with the prospect of an Allied invasion of Europe that their attitude changed. In many ways it mirrored that of the Japanese and was a tacit admission that their naval strategy had failed. German midgets were weapons of desperation, based on the hope that if used in sufficient numbers they would interrupt the Allied cross-channel supply lines. With the exception of the excellent *Seehund* two-man submarine, German midgets were poorly constructed and most were as lethal to their crews as they were intended to be to the opposition.

The operations of the various midget submarines during the Second World War remain some of the most supreme examples of cold-blooded courage in history. In a war which became dominated by technology and weapons of mass destruction, the achievements of the midget submariners of all countries stand out, harking back to an earlier and more honourable age, where individual courage and skill-at-arms were the attributes which won wars.

The post-1945 period has seen the midget submarine occupy an ambivalent place in the world's navies' order of battle. Britain and

The interior of the control-room of a British X-Craft midget submarine showing the OOW at the periscope. (Author)

America professed an initial interest in such craft in the immediate postwar period, but it was soon abandoned. The midget submarine was then taken up by some of the world's smaller navies: those of Yugoslavia, Colombia and Pakistan to name but a few. Other countries, such as North Korea, Libya and Iran, have also invested in midget submarines and their possession of these craft, in view of the unsavoury nature of these regimes and their predilection for supporting international terrorism, must be viewed with the greatest suspicion. The former USSR/Russia was always, and remains, a major player in the midget submarine field, and given the chronic cash shortage affecting all aspects of the Russian economy, the sale of Russian technology in this field to the highest bidder must be accepted. The USA abandoned midget submarine development after the failure of the X-1 craft in favour of concentration on the carrier, SSBN and SSN programmes. The end of the Cold War now means that the US Navy must adopt a more flexible approach and prepare for more limited operations, possibly in shallow coastal waters where SSN deployment would be inappropriate. They are currently developing the ASDS (Advanced SEAL Delivery System), a midget submarine with a marked resemblance to a British X-Craft, which can carry mines or swimmers.

One aspect of midget submarine operations which has survived almost unaltered since the Second World War is a successor to the wartime *Maiale*

X.1, *the only midget submarine ever to be commissioned by the United States Navy. (US Navy)*

X.23 *returns to the depot ship*
Largs *on 6 June 1944 after*
spearheading the Normandy
landings. (Author)

and Chariots, now known as Swimmer Delivery Vehicles (SDV). These are now far more sophisticated than the weapons in which Greenland and Visintini went to war but the basic principles are the same. They can be carried by submarines – a number of former SSBNs 'disarmed' by the SALT and START agreements are enjoying a new lease of life as SDV carriers. When it proved impossible to replace their Polaris missiles with the larger Poseidon system, USS *Sam Houston* (SSBN 609) and *John Marshall* (SSBN 611) of the Ethan Allen-class were converted to carry 67 SEAL swimmers in double dry-deck shelters fixed to the casing. Conversion started in 1984 but both boats were decommissioned in

September and November 1991 respectively. They were replaced by *Kamehameha* (SSBN 642) and *James Polk* (SSBN 645) of the Lafayette-class. On the Soviet side, a number of Project 66 and A boats (better known in the west as Yankee SSBNs) have been refitted for other duties by having their central section (containing the sixteen SS-N-4 missiles) removed and the two 'ends' of the boat being welded together. Nearly all modern submarines have escape chambers which can easily double as exit–re-entry chambers for divers while some 'conventional' submarines, such as the British *Otus* and *Opossum*, are fitted with five-man exit–re-entry chambers built into their fins. Whatever the refinements made to the *matériel* in modern submarines, the principles are the same as those under which *Decima Mas* operated so successfully. There is nothing very new in the midget submarine world that has not been done before.

Midget submarines have three advantages. First, the advantage of surprise: they could choose the place and time for an attack; second, the absence of any form of harbour defence (boom defence has been all but forgotten as an aspect of naval operations since 1945); third, the operators of these craft are fanatics, whether they be communist functionaries or Muslim zealots. The belief that death in action will lead to a Marxist or Islamic nirvana will overcome a good deal of inadequacy in training. It only takes one midget to get through with her cargo for the mission to succeed. One can only speculate at the result of such an operation: a wrecked off-shore oil installation in the Persian Gulf; an American aircraft-carrier mined while at anchor in the Bay of Naples or a chemical/biological (or nuclear) device detonated in an Israeli harbour.

The most exciting and promising developments centre on the field of robotic, unmanned craft which would be rather like homing torpedoes but with a passive/active search capability of their own. The British company Scion has developed SPUR (Scion's Patrolling Undersea Robot) while the Americans have investigated SMSPs (Small Mobile Sensor Platforms) deployed from torpedo tubes. In March 1990 the American Defense Advanced Research Projects Agency (DARPA) announced the construction of two prototype Unmanned Undersea Vehicles (UUV). An SSN, USS *Memphis* (SSN 691), has been converted to act as an at-sea test-bed for advanced submarine technology including the launch and recovery of UUVs. The UUV is designed as a tactical system which can be deployed from submarines, surface ships or direct from the shore and can perform a number of functions including mine detection, underwater surveillance (including ASW), and communications. The key to the performance of these functions is advanced electronic systems which include 'artificial intelligence' algorithms that function in the same way as human thought processes. In order to guard against computer failure, the vehicle will employ three fully redundant computers which will employ a 'voting' approach to system management on board the craft. All three computers must 'agree' on how the craft is run – if only two agree then the craft will continue to operate but in a degraded mode.

The UUV is 36ft long and 44in in diameter. Considerable attention has been paid to reducing the overall size of the craft by using advanced technology and reducing the size of the powerplant. An internal pressure hull will house the mission payload, occupying a 5ft long section. This would consist of the appropriate software and components for surveillance, communications or mine detection duties. In the case of the latter, the package includes an ultra-thin cable containing a fibre-optic communications link required for the transmission of commands from a surface ship or submarine. The propulsion system, consisting of a 12hp electric motor, and a motor controller, will occupy the after 12ft of the vehicle. The motor is built to operate even when completely flooded with sea water. Bearings are fabricated with a special non-corrosive alloy, and the copper windings that carry power to the motors are impervious to wear. During normal operations, the internal volume of the motor is filled with oil in order to equalise pressure between the inside and the outside of the motor, permitting the use of a thinner and lighter housing. Unmanned vehicles such as SMSPs and UUVs would enable a relatively small number of SSNs to 'control' a large area of ocean through which enemy forces would have to pass. They also offer considerable advantages in the field of mine detection and electronic surveillance, and are capable of unlimited under-ice operations – a field denied to manned midget submarines.

A number of conclusions can be drawn about midget submarine operations which are extremely relevant today. First, thorough and realistic training is essential if they are to be successful. In the past, British

The British submarine L24 *surfaces amid the tangle of an anti-submarine boom during boom defence trials in 1940. Net obstacles were once a routine defence against submarine incursions but this art has largely been forgotten. The rise in midget submarine development since the Second World War may mean that this ancient art has to be rediscovered. (Author)*

and Italian operators were well trained, and this was reflected in their achievements, whereas German operators were flung into battle with hardly any training at all, and achieved little as a result. However, the ability of the fanatic to score one decisive hit at the cost of his own life cannot be ignored. Second, midget submarines can be built quickly, cheaply and in large numbers. Moreover they are extremely easy to hide. Third, almost any merchant ship or submarine can be adapted to carry a midget submarine. Fourth, no defences have ever stopped a midget submarine attack. They have been a hindrance and have deterred some attackers but a small number of the midgets have always got through. Modern bases are virtually defenceless against this form of attack, especially as wartime skills in the field of boom defence have long since disappeared. Faced by the midget threat, simple last-ditch measures such as spilling oil fuel on the surface of the water will render a small periscope useless. The 'green lobby' would hate it, but against a simple craft dependent on periscope observation of the target, it would be highly effective. Fifthly, one-man operated craft are doomed to failure. A man on his own has too much to do and loses heart. There should be at least two in the crew, and for an operation of any duration at least four are required.

Lastly, the quality of the operators, the 'human resources', is vital. Men best suited to midget submarine operations are those least likely to fit in with the routines of a peacetime navy. It is interesting to note from the British perspective that Australians, New Zealanders and South Africans, whose antipathy to the Naval Discipline Act was legendary, were extremely competent X-Craft personnel. The risks implicit in training for midget submarine operations mean that such men have to develop a team spirit unique to their formation – this is what distinguished *Decima Mas* from the rest of the *Regia Marina* and what distinguishes Russian special forces from the run-of-the-mill conscript.

In 1907 the eminent historian Arthur Thayer Mahan wrote the following in *From Sail to Steam*: 'It is now accepted with naval and military men who study their profession, that history supplies the raw material from which they are to draw their lessons, and reach their working conclusions. Its teachings are not, indeed, pedantic precedents, but they are the illustrations of living principles.'

There is no doubt that midget submarines are still a force to be reckoned with. The development of new technologies will make them even more effective.

SUBMARINE
MISCELLANY

I n any submarine history there are deeds and stories which fail to fit conveniently into a chronological or thematic narrative. These stories reflect the heroism and tragedy associated with those engaged in submarine operations. In studying submarine history and operations, such stories have gathered themselves in my files until their inclusion in the book became a necessity. This selection is my personal choice and I have tried to be as impartial and international as possible.

Undoubtedly, one of the unluckiest submariners of all time was the Italian Nazario Sauro. Sauro was a merchant marine officer and a narrative of the city of Trieste. As such he was a citizen of the Austro-Hungarian Empire, and liable for naval service in the *KuK Kriegsmarine*, although he was an ethnic Italian and spoke Italian as his mother tongue.

Giacinto Pullino's *crew march into captivity in Pola in August 1916. (Kriegsarchiv, Vienna)*

When Italy declared war on Austria–Hungary in 1915, Sauro, like many others, fled across the border to offer his services to the Italian cause. Sauro was commissioned in the *Regia Marina* and appointed as a navigator to the submarine *Giacinto Pullino* where it was hoped that his specialist local knowledge could be best employed. On 30 July 1916 *Pullino* left Venice for a patrol in the Gulf of Fiume but while *en route* to her billet went aground off Galiola Island. She was abandoned by her crew after they had comprehensively wrecked the inside of the submarine, The Italians were later picked up by the Austrians and Sauro, now fearful for his life, tried to pass himself off as one Niccolo Sambo. However, the Austrians were suspicious: Sauro had been found swimming on his own, away from the main body of *Pullino*'s survivors with his face covered in bruises. Under interrogation some of *Pullino*'s crew revealed his true identity and hinted that they thought that Sauro had deliberately run the boat aground in order to ingratiate himself with his former masters. The whole affair is mystifying. Was the grounding of the *Pullino* an error or did Sauro deliberately run her aground? Had Sauro been beaten up by his 'comrades' after the grounding or were his injuries accidental ones sustained in climbing out of the boat?

Whatever the truth the Austrians put Sauro, still stoutly claiming his real name as Sambo, on trial on a charge of treason. Sauro maintained his innocence but his defence collapsed when his mother was brought into the court to identify him. The court found him guilty, sentenced him to death and he was hanged at Pola on 10 August 1916. Unlucky Sauro became Italy's first hero of the First World War. Perhaps more appropriately, in 1976 a new Italian submarine was launched which bears his name – she is still in service.

The voluntary self-sacrifice of life in battle has a long and honoured place in western military tradition. This is distinct from the spirit of the Japanese *kamikaze* who deliberately sought death. The western tradition is one of voluntary sacrifice of one's life to prevent capture or in the face of a hopeless situation . . . *Dulce et decorum est, pro patria mori*.[1] The history of submarine warfare provides numerous such examples.

On the night of 18 November 1943 USS *Sculpin* (Lt Cdr Fred Connaway USN) was attacking a Japanese convoy *en route* from Truk to the Marshall Islands. Connaway had been put down by the escort, had gone deep and let the convoy go over him and then had come to the surface to make an 'end around' the convoy to prepare for another attack. As *Sculpin* came to the surface the depth gauge stuck at 125ft. The diving officer, Ensign W. Fielder USNR, was unaware of this failure, so kept driving the boat up and pumping out water. *Sculpin* came to the surface in full view of the Japanese destroyer *Yamagumo* which had been left behind by the escort command as a 'sleeper'. As Connaway hurriedly dived the submarine, an eighteen-charge pattern exploded around her, doing considerable damage. Connaway decided that he had no alternative but to surface and try to fight it out. It was a one-sided engagement. *Sculpin*'s bridge was soon disabled and Connaway, the executive officer and the gunnery officer were killed. Command passed to Lt C.E. Brown USNR who ordered 'Abandon Ship'. The crew came up on to the casing as the Chief of the Boat opened the main vents.

The unfortunate Nazario Sauro, navigator of the Italian submarine Giacinto Pullino, *who was hanged by the Austrians in August 1916. The injuries on his face are clearly visible, but how he got them is a mystery. (Kriegsarchiv, Vienna)*

Capt John Cromwell USN was a passenger on board *Sculpin*. He was there to command a wolf-pack if *Sculpin*'s operations were coordinated with other submarines on patrol off Truk. Cromwell was fully cogniscent of American code-breaking activities and of the plans for the forthcoming invasion of the Gilbert Islands: Operation Galvanic. Cromwell knew that if he was captured, the Japanese would very likely realise that his presence on the submarine was unusual (even the Japanese would know that an American submarine was not commanded by a four-ring captain)

Lieutenant de Vaisseau *Roland Morillot, the self-sacrificing commanding officer of the French submarine* Monge, *sunk in December 1915. (Service Historique de la Marine, Paris)*

and torture him. Cromwell was scared that he might give away not only details of the plans for Galvanic but also of American code-breaking. Accordingly as *Sculpin* sank beneath the waves, Cromwell deliberately remained in the control-room. 'I know too much,' he shouted to Brown when the latter asked him to save himself. Several others from *Sculpin*'s ship's company elected to remain in their boat on her last and deepest dive. Cronmwell's fate was not known until after the war but when the details of his cold-blooded self-sacrifice became known, he was posthumously awarded the Medal of Honor, the second US submariner to be thus honoured.

The French submarine *Surcouf* has long been an object of fascination for the historian, not solely because of her unique size and armament (see page 77) but also because of the persistent rumours surrounding the loyalty, or otherwise, of her crew to the Allied cause during the Second World War. *Surcouf* was a large cruiser submarine built in 1929 with a displacement of 3,250 tons. She was designed for commerce warfare, the *guerre de corse* which had long been a central element of French naval strategy, and was armed with two 8in guns in addition to her more conventional armament of eight 21in and four 15.7in torpedo tubes. For

Defeat, 1943. An Italian Bronzo-class submarine lies interned at Malta. A British guard stands on the forward casing. (Author)

long-range reconnaissance she carried a Besson MB411 float plane in a watertight hangar on the casing. When the Germans invaded France in 1940 *Surcouf* was in refit at Brest but managed to escape to Devonport. There she was taken over by the Royal Navy in Operation Catapult in July 1940: two British personnel were killed in the process. The British, faced with the problem of what to do with this monster submarine, felt that she should be moored somewhere out of the way and forgotten about. Her machinery and armament were incompatible with British equipment and there was no hope of getting spares from France. De Gaulle would have none of it. *Surcouf* – the largest submarine in the world – was a symbol of French greatness.

So began a miserable saga in which the British tried to keep the submarine operational while her crew, drafted from such French naval personnel who opted to join the Free French cause, grew steadily more disenchanted. Rumours about her loyalty or otherwise multiplied.[2] In the end Vice Adm Horton decided to send her off to the Pacific to guard French possessions in Polynesia, thereby killing two birds with one stone: satisfying the French that *Surcouf* was still useful and getting rid of her to another theatre of operations.

Accordingly, *Surcouf* sailed from Halifax, Nova Scotia, via Bermuda and the Panama Canal to Tahiti and New Caledonia. Until recently, it has always been believed that she was sunk in collision with the American freighter *Thompson Lykes* on 18 February 1942, from which there were no survivors. However, attention has now focused on the circumstances of the sinking and there is room for believing that *Surcouf* was not the victim of the collision. The evidence for this centres around the positions for the two ships at the time the collision occurred. *Surcouf*'s estimated position and that of the *Thompson Lykes* are about 50 miles apart, though this can easily be explained away as a navigational error. However, further doubt has been cast on the collision theory as a result of analysis of photographs showing damage to the bows of the *Thompson Lykes* taken after the collision. The energy expended by a mass of 6,000 tons (*Thompson Lykes*'s displacement) striking another while moving at 13 knots, her speed at the time of the collision, is more than sufficient to cause major repairs. The photographs show no such damage.[3] If *Thompson Lykes* did not strike *Surcouf*, then what did she collide with? Several members of her crew described the object as a launch of some description but this has always been discounted in view of the fact that if a launch or smaller tanker had been lost then her failure to return to port would have been noticed. However, the ship struck by the *Lykes* might have been a smuggler's launch – whose absence would not have been reported.

A few days after the collision, reports were heard of an attack on a submarine by two A-17 aircraft and a B-18 from the US Army Air Force's 3rd Bombardment Squadron flying from Rio Hata in the Panama Canal Zone. The orders to attack a submarine were received from 6th Bombardment Group HQ in Panama at 07.13 hrs on 19 February although a search of the documentary evidence has not established how 6th Bombardment Group HQ were alerted to the submarine's presence.[4] The attack took place in an area where *Surcouf* would have been had she continued along her course to Colon, at the eastern end of the Panama Canal. She was travelling in a moving haven but there is no record of the 3rd Bombardment Squadron being briefed about this before they took off – although all US commands in the area had been issued with details of her route. Lt Harold Staley was the pilot of one of the A-17s and recalled that the submarine, 'a very large one', was lying very low in the water and after Staley had dropped four 100lb bombs, she drifted towards the coast and sank off the island of Chi Chi Mey in the Archipelago de las Mulatas. Bodies were washed ashore and buried by local Indians near San Blas Point. Although the details of the attack are recorded in the operational records of the 3rd Bombardment Squadron, there are, unfortunately, no records of the operational orders from the 6th Bombardment Group, and no documents indicating how they learned of the submarine's presence. The end of the *Surcouf* is shrouded in mystery but there is considerable doubt over whether she was sunk by collision. On the other hand there is insufficient evidence to say categorically that she was the victim of Friendly Fire. As she did in life, *Surcouf* continues to perplex and confuse the historian.

When Germany invaded Poland on 1 September 1939, the small Polish Navy was all but overwhelmed. A number of units managed to escape from the Baltic to England, including the submarines *Orzel* (Cdr J. Grudzinski)

The Polish submarine Orzel *(left) at Rosyth after her dramatic escape from the Baltic in 1939. (Author)*

and *Wilk* (Lt Cdr Krawczyk) ORP. The Polish submarine *Orzel* was launched in January 1938 at Vlissingen. Her completion was pushed through quickly for fear of German sabotage and she arrived at Gdynia in April 1939. Her first war patrol was terminated by mechanical problems and she sought shelter in Tallinn (Estonia) on 15 September where she was interned. Charts any navigational instruments were removed but on 18 September *Orzel* escaped and arrived at Rosyth on 14 October. The voyage had been made without any navigational instruments and evading German minefields and air patrols. *Orzel* was subsequently lost on or around 8 June 1940 in the North Sea – probably as a result of hitting a mine.

NOTES

1   'It is sweet and fitting to die for one's country', attrib. Wilfrid Owen.
2   For a full account of *Surcouf*'s short life see Richard Compton-Hall, *Monsters and Midgets* (Blandford Press, 1983) and James Rusbridger, *Who Sank* Surcouf? (Century Random House, 1991).
3   Imperial War Museum, Department of Photographs, correspondence with James Rusbridger, RUSB/J/PH.
4   Author's correspondence with Office of Air Force History, Maxwell Air Force Base.

BIBLIOGRAPHY

Adams, T.A. and Lees, D.J. *Register of Type VII U-boats* (London, World Ship Society, 1991).

Admiralty. *Preliminary Narrative of the War at Sea* (BR 1738), 6 vols (London, 1944 onwards).

——. *German, Italian and Japanese U-boat Casualties during the War* (London, HMSO, 1946).

——. *German Warships of World War One* (London, Greenhill Books, 1992).

——. Naval Staff History of the Second World War, *Submarines Vol. 1 Operations in Home, Northern and Atlantic Waters* (London, 1953).

——. Naval Staff History of the Second World War, *Submarines Vol. 2 Operations in the Mediterranean* (London, 1955).

——. Naval Staff History of the Second World War, *Submarines Vol. 3 Operations in Far Eastern Waters* (London, 1956).

——. Naval Staff History of the Second World War, *Battle Summary No. 22, Arctic Convoys 1941–1945* (CB 3305[4]) (London, 1954).

Aichelburg, W. *Die Unterseeboote Osterreich-Ungarns*, 2 vols (Graz, Akademische Druck u Verlagsanstalt, 1981).

Bacon, Adm Sir Reginald, *The Dover Patrol*, 2 vols (London, Hutchinson, 1919).

Bagnasco, E. and Spertini, M. *I Mezzi D'Assalto Della X^a Flotiglia MAS, 1940–1945* (Milan, Albertelli Editore, 1991).

Barnett, C. *Engage the Enemy More Closely – The Royal Navy in the Second World War* (London, Hodder & Stoughton, 1991).

Beesely, P. *Very Special Intelligence* (London, Hamish Hamilton, 1977).

Belknap, R.R. *The Yankee Mining Squadron* (USNI, Annapolis, 1920).

Burn, A. *The Fighting Captain* (London, Leo Cooper, 1993).

Campbell, G. *My Mystery Ships* (London, Hodder & Stoughton, 1928).

Caulfield, Max. *The Easter Rebellion* (Dublin, Gill & MacMillan, 1995).

Chalmers, W.S. *Max Horton and Western Approaches* (London, Hodder & Stoughton, 1954).

Chatterton, E.K. *Beating the U-boats* (London, Hurst & Blackett, 1943).

Compton-Hall, R. *The Underwater War 1939–1945* (Poole, Blandford Press, 1982).

—— and Moore, J.E. *Submarine Warfare Today and Tomorrow* (London, Michael Joseph, 1983).

——. *Submarine vs Submarine – The Tactics and Technology of Underwater Confrontation* (London, Grub Street, 1988).

——. *Submarines and the War at Sea* (London, Macmillan, 1991).

Copiero del Villar, Jesús Ramirez. *Espias y Neutrales: Huelva en la II Guerra Mundial* (Huelva, Imprenta Jimenez, 1996).

Corbett, Sir Julian S. and Newbolt, Sir Henry. *History of the Great War, Naval Operations*, 5 vols (London, Longman, 1920–1931).

Cremer, P. *U-boat Commander – A Periscope View of the Battle of the Atlantic* (London, Bodley Head, 1984).

Dönitz, K. *10 Jahre und 20 Tage* (Frankfurt, Athenaeum, 1958).

Dorling, T. *Swept Channels* (London, P. Allen, 1935).

Edwards, K. *We Dive at Dawn* (London, Rich & Cowan, 1939).

Farago, Ladislas. *The Tenth Fleet* (New York, Obolensky, 1962).

Franks, N. *Conflict over the Bay* (London, Grub Street, 1986).

——. *Search, Find and Kill, The RAF's U-boat Successes in WW2* (London, Grub Street, 1995).

Friedman, Norman. *Desert Victory – The War for Kuwait* (Annapolis, US Naval Institute Press, 1991).

Gallery D.V. *Clear the Decks!* (New York, Warner, 1951).

——. *Twenty Million Tons Under the Sea* (Chicago, Regnery, 1956).

Gayer, A. *Die deutschen U-boote in ihrer Kriegfuhrung 1914–1918* (Berlin, Mittler, 1930).

Geise, O. and Wise, J.E. *Shooting the War – Memoirs of a WW2 U-boat Officer* (Annapolis, US Naval Institute Press, 1994).

Gibson, R.H. and Prendergast, M. *The German Submarine War 1914–1918* (London, Constable, 1931).

Goulter, Christina. *The Forgotten Offensive: Royal Air Force Coastal Command's Anti-Shipping Offensive, 1940–1945* (unpublished PhD thesis, University of London, 1993).

Grant. R.M. *Known Sunk – German Warship Losses 1914–1918* (US Naval Institute Proceedings, vol. 64 (1938)), pp. 66–77.

——. *U-boats Destroyed* (London, Putnam, 1964).

——. *U-boat Intelligence 1914–1918* (London, Putnam, 1969).

Gray, E. *Few Survived: A History of Submarine Disasters* (London, Leo Cooper, 1986).

Gretton, Vice Adm Sir Peter. *Convoy Escort Commander* (London, Cassell, 1964).

——. *Crisis Convoy* (London, Peter Davies, 1974).

Groener, E. *German Warships*, vol. 2 (London, Conway Maritime Press, 1991)

Hackmann, W. *Seek and Strike, Sonar, Anti Submarine Warfare and the Royal Navy, 1914–1954* (London, HMSO, 1984).

Hague, A. and Ruegg, R. *Convoys to Russia, 1941–1945* (London, World Ship Society, 1992).

Halpern, Paul. *The Naval War in the Mediterranean, 1914–1918* (London, Allen & Unwin, 1987).

Hashagen, Ernst. *The Log of a U-boat Commander* (London, Puttnam, 1931).

Hervey, Rear Adm J. *Submarines* (London, Brasseys, 1994).

Hessler, G. *The U-boat War in the Atlantic* (London, HMSO, 1989).

Hinsley, F.H., Thomas, E.E. *et al. British Intelligence in the Second World War – Its Influence on Strategy and Operations* (London HMSO, 1979–88).

—— and Stripp, A. *Codebreakers – The Inside Story of Bletchley Park* (Oxford, Oxford University Press, 1993).

Howse, D. *Radar at Sea*, 3 vols (London, Macmillan/Naval Radar Trust, 1993).

Huan, Claude. *Les Sous Marins Français, 1918–1945* (Paris, Marines Edition, 1995).

Hoy, H.C. *40 OB or How the War was Won* (London, Hutchinson, 1932).

James, Adm Sir William. *The Codebreakers of Room 40* (New York, Methuen, 1956).

Johnson, B. *The Secret War* (London, BBC Books, 1978).

Jones, G. *Submarine vs U-boat* (London, William Kimber, 1986).

Kahn, D. *Seizing the Enigma* (Boston, Houghton Mifflin, 1991).

Kemp, P. *The T-Class Submarine* (London, Arms & Armour Press, 1990).

——. *Friend or Foe, Friendly Fire at Sea during the Second World War* (London, Leo Cooper, 1995).

——. *U-boats Destroyed* (London, Arms & Armour Press, 1997).

—— and Wilson, M. *Mediterranean Submarines* (Manchester, Crecy Books, 1997).

Keyes, Adm Sir Roger. *Naval Memoirs Vol. 2, From Scapa Flow to the Dover Straits* (New York, Butterworth, 1935).

Lewin, R. *Ultra Goes to War* (London, Hutchinson, 1978).

Macintyre, D. *U-boat Killer* (London, Weidenfeld & Nicolson, 1958).

Mallmann Showell, J.P. *U-boat Command and the Battle of the Atlantic* (London, Conway Maritime Press, 1989).

Middlebrook, M. *Convoy, the Battle for SC.122 and HX.229* (London, Allen Lane, 1976).

Milner, M. *North Atlantic Run, The Royal Canadian Navy and the Battle for the Convoys* (Annapolis, US Naval Institute Press, 1985).

——. *The U-boat Hunters* (Annapolis, US Naval Institute Press, 1994).

Ministry of Defence (Navy). *British Mining Operations 1939–1945*, BR1736(56)(1) (London 1973).

—— (Ship Department). *The Development of HM Submarines*, BR3043 (January 1979).

Morison, S.E. *History of US Naval Operations in World War II, The Atlantic Battle won May 1943–May 1945* (Boston, Little Brown, 1990).

Morsier, P. de. *Les corvettes de France libre* (Paris, Ed. France–Empire, 1972).

Padfield, P. *Dönitz, The Last Führer* (London, Gollancz, 1984).

——. *War Beneath the Sea* (London, John Murray, 1995).

Poolman, K. *HMS Vindex* (London, William Kimber, 1983).

Price, A. *Aircraft vs Submarine* (London, William Kimber, 1973).

Robertson, T. *The Golden Horseshoe* (London, Evans Brothers, 1955).

——. *Walker RN* (London, Evans Brothers, 1956).

Rohwer, J. *The Critical Convoy Battles of March 1943* (London, Ian Allen, 1977).

——. *Axis Submarine Successes 1939–1945* (Annapolis, US Naval Institute Press, 1983).

—— and Hummelchen, G. *Chronology of the War at Sea* (London, Greenhill Books, 1992).

Roscoe, T. *United States Submarine Operations in World War II* (Annapolis, US Naval Institute Press, 1988).

Roskill, S.W. *The Secret Capture: The Story of U110* (London, Collins, 1959).

——. *The War at Sea 1939–1945*, 4 vols (London, HMSO, 1954–1961).

Rossler, E. *The U-boat: The Evolution and Technical History of German Submarines* (London, Arms & Armour Press, 1981).

Runyan, T. and Copes, J.M. (eds). *To Die Gallantly – The Battle of the Atlantic* (San Francisco, Boulder Press, 1994).

Sainsbury, A. and Shrubb, R. *The Royal Navy Day by Day* (Centaur Press, 1979).

Schaeffer, H. *U-boat 977* (London, William Kimber, 1952).

Schoenfeld, M. *Stalking the U-boat – USAAF Offensive Anti-Submarine Operations in WW2* (Washington, Smithsonian, 1995).

Schull, J. *The Far Distant Ships* (Ottawa, Ministry of National Defence, 1961).

Shelford, W.O. *Subsunk: The Story of Submarine Escape* (London, Harrap, 1960).

Simpson, G.W.G. *Periscope View* (London, Macmillan, 1972).

Sokol, H.H. *Osterreich Ungarns Seekrieg 1914–1918*, 2 vols (Graz, Akademische Druck u Verlagsanstalt, 1967).

Spindler, Adm Arno. *Der Krieg zur See 1914–1918, Der Handelskrieg mit U-booten*, 5 vols (Berlin, Mittler, 1932–66).

Stern, R. *Type VII U-boats* (London, Arms & Armour Press, 1991).

Sternhall, C.M. and Thorndike, A.M. *Anti-submarine Warfare in World War Two, Operational Evaluation Group Report No. 51* (Washington, 1946).

Syrett, D. *The Defeat of the German U-boats* (Columbia SC, University of South Carolina Press, 1994).

Tarrant, V.E. *The U-boat Offensive 1914–1945* (London, Arms & Armour Press, 1989).

——. *The Last Year of the Kriegsmarine* (London, Arms & Armour Press, 1994).

Terraine, J. *Business in Great Waters, The U-boat Wars 1916–1945* (London, Leo Cooper, 1989).

Thetford, O. *British Naval Aircraft since 1912* (London, Puttnam 1977).

Thompson, J. *The Imperial War Museum Book of the War at Sea* (London, Sidgwick & Jackson, 1996).

Ufficio Storico della Marina Militare: La Marina Italiana Nella Seconda Guerra Mondiale. Vol. XIII, *I Sommergibile in Mediterraneo, Dal 10 Guigno 1940 al 31 Decembre 1941* (Rome, 1972).

Watts, A. *The U-boat Hunters* (London, Macdonald & Janes, 1976).

Wemyss, D.E.G. *Relentless Pursuit* (London, William Kimber, 1955).

Whinney, R. *The U-boat Peril* (London, Blandford, 1987).

Wingate, J. *The Fighting Tenth* (London, Leo Cooper, 1991).

——. *The Forgotten Fleet* (London, Michael Joseph, 1969).

——. *Ultra at Sea* (London, Leo Cooper, 1988).

Woodman, R. *Arctic Convoys* (London, John Murray, 1993).

Y'Blood, W.T. *Hunter Killer: US Escort Carriers in the Battle of the Atlantic* (Annapolis, US Naval Institute Press, 1983).

Young, E. *One of Our Submarines* (London, Hart Davis, 1953).

INDEX